T0369327

Also by John Zumerchik

Encyclopedia of Sports Science
(Macmillan Library Reference USA) 2 v., 1997. Editor

Macmillan Encyclopedia of Energy
(Macmillan Reference USA) 3 v., 2001. Editor-in-Chief

John Zumerchik

SIMON & SCHUSTER PAPERBACKS

NEW YORK LONDON TORONTO SYDNEY

Newton on the Tee

A Good Walk Through the Science of Golf

SIMON & SCHUSTER PAPERBACKS
A Division of Simon & Schuster, Inc.
1230 Avenue of the Americas
New York, NY 10020

First Simon & Schuster trade paperback edition April 2008

SIMON & SCHUSTER PAPERBACKS and colophon are registered trademarks of Simon & Schuster, Inc.

For information about special discounts for bulk purchases, please contact Simon & Schuster Special Sales at 1-800-456-6798 or business@simonandschuster.com

Designed by Karolina Harris

Manufactured in the United States of America

10 9 8 7 6 5 4 3 2 1

The Library of Congress has cataloged the hardcover edition as follows:
Zumerchik, John.
Newton on the tee : a good walk through the science of golf / John Zumerchik.
p. cm.
1. Swing (Golf). 2. Golf—Physiological aspects. 3. Golf—Equipment and supplies—Miscellanea. 4. Golf—Miscellanea. 5. Sports sciences. I. Title.

GV979.S9 Z86 2002
796.3523—dc21 49711049

ISBN-13:978-1-4165-4129-5

Acknowledgments

As with any work of this nature, I have many people to thank. This book would not have been possible without the support of truly brilliant individuals who could answer my questions. In the motor control area, Tommy Boone offered valuable suggestions on the mind-body aspects of the game, and Don Hopkins, O'Dean Judd, Don Lichtenberg, David Lind, and Arjun Tan all offered valuable suggestions in reviewing the chapters covering physics. In particular, Don Hopkins provided extensive help with the physics of putting. Drawing on the research that he had done on the physics of bowling, he developed models that made it possible to explain the very complex dynamics involved in putting.

I also want to thank the individuals who helped convert my rough manuscript into a polished book: Gustavo Camps for his assistance with the illustrations; Jon Malki and Jeff Neuman, my editors at Simon & Schuster; Red Wassenich, my copyeditor; and Karen Kim, my proofreader.

John Zumerchik

Simon & Schuster Paperbacks

New York London Toronto Sydney

This book is dedicated to John and Eileen, my parents, for their patience and guidance, to my wife, Ana, for her love and support, and to my three children, Andre, Janine, and Giselle, for the great joy and inspiration that they bring to my life.

Contents

Preface

Newton on the Tee describes, in simple everyday language, the most interesting and useful physics and motor control research applicable to golf. The genesis for this book occurred as I was thumbing through articles in preparing the two-volume *Encyclopedia of Sports Science,* and it occurred to me that it would be worthwhile to take the little-known general research that has been done on the physics and motor control aspects of sports and apply it specifically to golf. In writing for the typical golfer, my foremost concern was trying to present the scientific principles without excessive use of technical terminology or golf jargon. To help bridge the chasm between the languages of golf and science, I present parallels to anecdotal comments from professional golfers that explain scientific principles in a descriptive but unscientific fashion and also use analogies to actions in other sports, because the forces that are subtle and invisible in golf are often obvious and easy to picture when appearing in other sports. These analogies should be especially helpful for natural athletes who have displayed talent in a number of other sports, yet have found golf exceedingly difficult.

Regardless of ability, golf is an intoxicating lifelong passion for many because of a natural desire to compete and improve. Golfers forsake family and friends and each year spend billions of dollars searching for tips in books and magazines, picking up pointers from instructional videos, seeking customized help from local professionals, or shopping for miraculous swing-improving equipment that promises to shave strokes from every user's scorecard. All is done in an effort to drive the ball a little farther, pitch the ball with greater bite, and drop long putts with more frequency. In this quest, you can either change your swing, change what you're swinging, or change the swinger. *Newton on the Tee* tries to give you insight into all three approaches. Of course, in our modern world of instant gratification, the latter—improving one's physical fitness—is usually the option of last resort. It is much easier to seek real or imagined relief with a "quick fix" instructional lesson or a new driver than to suffer through the agony of improving your flexibility or replacing twenty pounds of fat with twenty pounds of muscle mass.

Only a few golfers have the interest and time to embark on a quest for mental and physical perfection. Therefore, *Newton on the Tee* is organized so that you can quickly find what interests you most. Chapter 1, "How Easy Is this Game Anyway?" begins by describing why golf is a much tougher game than it looks, detailing all the little things that you have to get right to play well. In Chapter 2, "The Physics of a Sweet Swing," and Chapter 3, "Mind over Muscle," the focus is on the universal principles shared by all those very rhythmic and well-timed golf swings you can see on television every weekend—constituting the most vital aspects of the physical and mental game that need to be conquered in your quest for swing perfection. Chapter 4, "Getting the Ball from Here to There," covers shot-making decisions in all types of wind, weather and course conditions. Chapter 5, "Clubs and Balls," follows by describing the role of golf gear: the tradeoffs inherent in material choices, the pros and cons of club designs, and why the clubs and balls of today are far superior to those of yesteryear.

Since golf is a lifetime game, Chapter 6, "Injuries and Aging," spells out some of the physics and physiology behind aches and pains and what can be done to prevent injuries and to delay the age-related problems that cause a decline in play. And Chapter 7, "Probability and Statistics," finishes with a look at interesting mathematical applications that will enhance your understanding of the game of golf. "Call out boxes" can be found in most chapters that look into noteworthy scientific questions and explanations that are tangential to the material being presented.

People look to science for profound truths about the world around us, and with good reason. Scientific discoveries have led to the development of some truly amazing technologies in almost every imaginable field of study. Nevertheless, if you're looking for a long list of profound discoveries in *Newton on the Tee*, you're going to be disappointed. Science has actually contributed very little to the discoveries made in golf: It usually only confirms what has been proven through the endless trial and error experimentation that takes place every day on golf courses throughout the world. The history of innovations in golf often goes something like this: An experiment is tried, results follow that show it works, and then only later, after the innovation has been widely adopted, is it proven to be scientifically sound. For instance, up until very recently the design of equipment, selection of materials, and evaluation of performance have been rather scientifically unsophisticated. Usually this haphazard analysis has been used primarily to confirm the efficacy of an existing direction of innovation or an improvement already advocated; rarely has scientific research alone resulted in a revolutionary discovery. The greatest contribution of science has been in the refinement of innovations after a scientific analysis has shown that it works and why, which is my focus in this book. My objective is to clearly, concisely, and accurately present the scientific principles of golf to heighten your appreciation of the game and, with this knowledge, perhaps even improve the quality of your play.

Newton on the Tee

1

How Easy Is this Game Anyway?

THE intriguing paradox of golf is that almost everyone can play the game but oh so few can master it.

Since anyone can play the game, conventional wisdom is that golf is easy. Many spectators would concur. From watching PGA or LPGA golfers compete, it is legitimately believed by many viewers that they possess greater overall athletic talent, and therefore, if given the chance to relive life, one spent working on their game as much as the pros do, they too could have competed against the very best. This belief is even more pervasive among professional athletes competing in the other major sports. Many professional athletes are under the delusion that once they retire from playing their chosen sport, they have the ability to reach the top ranks of the PGA Tour because of their sheer athleticism. Although many have tried, it has never happened and probably never will. A few exceptionally talented athletes have been able to excel at professional baseball and football, or track and football, but no athlete has graduated to the PGA Tour.

No matter how talented the athlete, upper-echelon golf requires far too much precision to convert a hobby into a professional livelihood. Those that try eventually realize that it is too late to be great—their nervous systems are too hard-wired to match the fine-tuned precision found among golfers on the PGA Tour. Most of these would-be golf pros manage to prop up their egos by reminding themselves that if they concentrated on a PGA career it could have been a reality. Perhaps that's true, but you can imagine how envious other ballplayers are of golf pros: careers of most professional athletes average in the five-to-ten-year range while the careers of some golfers can span decades. Sam Snead actually won PGA tournaments in seven different decades-the 1930s, '40s, '50s, and '60s on the PGA Tour, and the 70s and '80s on the Senior PGA Tour.

Athletes with ample physical gifts often miscalculate their ability to excel at golf because the enormous talents displayed on tour are difficult for the untrained eye to pick up; they are either extremely subtle or completely invisible. Athletes who have been successful at every endeavor they have ever tried cannot achieve the same success golfing. Whereas being tall or possessing an athletic build are significant benefits in most sports, they're not in golf since the game does not feature explosive running and jumping or feats of great power, strength, or stamina. Power, speed, and jumping ability are very visible talents; golf talent, on the other hand, is not so apparent. Aside from the athleticism to swing the golf club with great rhythm and timing—an ability to subtly control hundreds of muscles—it is equally important that professional golfers have the mental skills and the game savvy to best take advantage of their physical skills.

The Precision for the Collision

In golf, looks can indeed be deceiving. From your living room recliner, golf looks easy. The same basic swing, with a few minor variations, seems to work for the longest drives to the shortest wedge shots. However, by just watching televised tournaments it is impossible to ever develop an appreciation for the thousands of subtle muscle movements involved in a swing that takes a little less than a second from start to finish and a collision that is over in 0.0005 of a second. And these movements must be held in memory, performed in the right sequential order, and carried out with precise timing. That is why the quest for a graceful, fluid, repeatable swing is an extremely humbling exercise for mind and body.

It's easy to tell someone to watch the pros and try to emulate their every move, but these are truly gifted athletes who work their magic on the ball in ever so subtle ways. We're not endowed with the abilities of Tiger Woods or Greg Norman; we don't have the same multifaceted skills, the same ability to employ invisible and subtle forces to alter the ball's flight, bounce, and roll. Thus, it's not worth trying to emulate their every move. Instead you want to honestly evaluate your strengths and weaknesses before selecting equipment and devising a strategy for play that best fits with your unique God-given talents. There is not a one-size-fits-all "right" swing. You need to develop a swing that plays up your strengths and deemphasizes your weaknesses.

Another aspect of the game that makes golf so difficult is that each collision does not have the same goal; the game plan and situation dictate not only what clubs should be used, but also adjustments in swing mechanics. With any given club, there can be a range of objectives: maximizing or minimizing impact velocity, trajectory, or spin to name a few. There are hundreds of different combinations of initial velocities, trajectories, and rates of ball spin that can get the ball in the proximity of where you want it to go; therefore, you have to approach shot-making like a statistician, figuring out all the probable outcomes from the many different trajectories that can get the ball near the hole. Then you have to not only select an option that offers good results with a comfortable margin for error, but also it has to be an option that you have confidence in your ability to execute. This optimal combination obviously will differ between hackers and pros (the pro has a much wider range of skills to execute those hundreds of combinations), but it is likely to vary between pros as well. That is one of the reasons it is so intriguing to watch the shot-making decisions of professional golfers. For those who know the game well, the chosen combinations of velocity, arc, and spin are as eagerly anticipated as the actual results.

Even when the goal is the very simple and straightforward one of maximizing impact velocity, there are thousands of things that can go wrong. If the clubface is a half-degree less than square to the ball, the ball can go off course by 20 yards—endangering the lives of unsuspecting fellow golfers as it slices its way across an adjacent fairway. If the downswing is just the slightest bit serpentine, the ball can slice 30 yards off-course and travel a much shorter distance. If club velocity is 110 miles per hour and not 111, it could be the difference between the ball landing a few feet from the hole or rolling right off the green. And if the ball is struck one-fourth inch too high or too low, the ball can, respectively, weakly dribble its way up the fairway or sky and land very short.

The ball's lie can also pose considerable problems, significantly altering the dynamics of the collision. Golf would be a whole lot simpler if you could take a drop until happy with the lie. Unfortunately, errant shots are likely to be in poor positions that lead to bad lies, which means there is going to be a lot of interference with your clubface's pathway to the ball, and if buried in the rough, a lot of grass sandwiched between your clubface and the ball at contact. The degree of difficulty gets magnified even more for lies in difficult terrain. Side-hill lies or balls coming to rest atop a ridge or in a ditch are fiercely difficult shots, because the difference in elevation of the ball in relation to your feet alters the distance from your hands to the ball, making the effective length of your club either too short or too long. You must make adjustments to compensate for this length difference: With an uphill lie, you have to adjust for the tendency to pull your shot, and with a downhill lie, the tendency for the ball to slice.

Wind, especially when gusty, adds still another layer of complexity. The dimpled golf ball might be an aerodynamic wonder that minimizes the effects of the wind, yet the wind can still drastically and unpredictably alter the flight of a golf ball since it is light and stays aloft for several seconds. How much the wind veers ball flight not only depends on its direction and velocity, but also on the ball's velocity, trajectory, time aloft, and the amount of backspin. If a 120-yard approach shot is taken expecting a 20-mph headwind, and it turns out to be 30, the ball might drop 20 yards in front of the green instead of at the pin. This is where wind strategies come into play, such as using a higher velocity and lower trajectory shot that will be less affected by the wind than a high-arcing shot, since the ball will be aloft for a shorter period and wind velocities are slower closer to the ground. The tradeoff, however, is that a higher velocity and lower trajectory shot will be tougher to stop as quickly, yet it can be well worthwhile since it allows for greater confidence in directional precision.

Precision shot-making requires plenty of analysis; demanding coordination of the physical, mental, and motor control aspects of the game; and an astute evaluation of how the environment is going to affect your game on any given day. Further, it requires an understanding of your equipment and how to get the most out of it under any circumstances. Athleticism helps develop finely tuned ball-striking skills but is of little consequence in cultivating the ability to confidently evaluate the environmental conditions and, considering your abilities, select a shot strategy that results in the best probable outcome.

Putting It All Together and Keeping It There

There is so much that needs to be put together to win at the elite level. Then once the magical combination is reached, many professionals will tell you how devilishly difficult it is to keep it together. For no apparent reason, the ability to execute sometimes just leaves a golfer. Injuries are sometimes to blame, but just as often, going into a funk happens for no apparent reason. There seems to be no easy explanation for why a professional golfer turns in an outstanding performance one week and proceeds to fall apart the next. Mark Calcavecchia broke a 46-year-old 72-hole scoring record with a 256 at the 2001 Phoenix Open. You would think a golfer that hot would be the favorite to win the next week, but four days later, he recorded a 40 on the front nine at the AT&T Pebble Beach Classic and wound up missing the cut.

A more routine phenomenon than the sudden and devastating slump is the prolonged losing streak that can be experienced by even the most talented golfers. Davis Love III was at the top of his game in winning the PGA Championship at Winged Foot in 1997, but then went on a long dry spell—playing the next 62 tournaments over 34 months without a victory. Until his seven-stroke comeback to win at Pebble Beach in February 2001, he was beginning to be referred to as the most talented golfer not capable of winning. Mark O'Meara had a career year in 1998, winning two Majors and being named the PGA player of the year, but he has not won a tournament since. It might take years for the precision skills of professional golfers to finally come together, and then once it does, success can be fleeting.

All golfers will attest to the difficulty of maintaining the groove. Professional golfers go through perhaps more ups and downs than any other athletes, and in addition, the human body is always changing, which impacts a golfer's game. It is nearly impossible to maintain the same quickness, flexibility, strength, and mental sharpness day to day, week to week, and month to month, which makes it extremely difficult to keep a hot streak going. And the changes are more significant than just the battle of the bulge, the occasional gaining or losing of a couple of pounds; the skeletal structure itself isn't static. Though bone length is fixed by the late teens or early twenties, width and density peak in the 30 to 35 range and then decline. Posture is also likely to change due to gravity; if you have bad posture, gravity continues to make it worse, slowly altering your swing dynamics. And finally, your flexibility isn't static. It improves if you work at it, deteriorates if you don't, and is prone to decline with age. Any one of these physical changes has the potential to change your swing in little ways that could have minor or major repercussions. Moreover, your brain complicates matters by sending out motor signals to the body that must be altered to account for changes, be it injury, aging, or just day-to-day physical well-being differences. This includes the doubt that keeps creeping into your head about whether these adjustments are right.

There is also the genetic component—the predisposition to injury and the effects of aging. Your body has to hold up. Regardless of your natural athletic talents, it takes years of practice to master the game. An injury at any time can end what could have been a very promising career. Top professional golfers have to constantly adjust their game for the injuries and pain that usually accompanies aging. The first adjustment is to the injury, followed by the adaptation back again after recovery. As motor memory switches from the healthy swing to the injured swing and then back again, there will always be a need to reestablish confidence.

Considering all the myriad factors that go into a sweet and confident swing, it is no wonder professional golfers are unable to articulate the reason for a slump. So the next time you hear someone say golf is an easy game, rest assured that you are listening to someone who has never considered how difficult it is to play well and continue to do so, and how little it takes to start playing poorly.

2
The Physics of a Sweet Swing

PROS drive the ball close to 300 yards and puncture greens with lawn-dartlike approach shots. Most of us can only dream about such heroics. We're elated when our shots carry over 200 yards and usually have to pray for lucky bounces and kind rolls to aid errant approach shots. There are no easy answers to why their swings can look so balanced and ours so unstable, theirs so effortless and ours so labored, theirs so smooth and ours so jerky. The only certainty is that pros better know how to fully exploit their God-given talents and equipment. From years of practice and continual fine-tuning, they have learned how to generate tremendous clubhead velocity to drive the ball vast distances while also developing the feel and control to achieve pinpoint accuracy for shot-making. Much of the difference between the duffer's hack and the pro's swing seem subtle; however, it is actually several subtleties, largely imperceptible to all but the best-trained eyes, that constitute the disparity. Blurring things further is the fact that no two golfers swing exactly alike, even professionals. Everyone possesses unique physical characteristics (height, weight, and frame) and athletic skills (speed, quickness, strength, and flexibility), rendering moot the notion of one universally identifiable "correct" perfect swing to be working toward. Differences in mechanics should be expected, and length of backswing, degree of wrist cock, plane of the swing, type of stance, and grip should all be tailored to your strengths, weaknesses, and idiosyncrasies. For example, if you have an athletic build—thin waist and wide shoulders—your best potential swing will look far different than a golfer built like a barrel. Likewise, if you are strong but lack flexibility, your best potential swing will be far different than a long, lanky golfer with great flexibility.

Although swing perfection is a largely unrealistic goal, a more attainable worthwhile one is developing a sweet swing based on a few fundamental principles of physics. The biggest hurdle to overcome in achieving this goal is developing the skill to control the thousands of well-timed muscle actions that go into a sweet swing. Pros perform these actions almost subconsciously, but for the rest of us (that is, everybody who did not start swinging a golf club as a toddler), the development of a sweet swing requires a much more conscious effort.

Power: Where It Comes from and How to Unleash It

Swing power is the most coveted trait in golf. If you have any doubts, camp out at a driving range for an afternoon: You'll see a disproportionate amount of time spent swinging drivers. Perhaps it is the simple thrill of seeing the ball sail a long way, the best way to release everyday stress, or perhaps it's just the macho thing to do. Whatever it may be, there is a love of power and a yearning to achieve more of it.

Power is very seductive for the fan as well. When a pro swings ferociously, "oohs" and "aahs" ring out from galleries. If a golfer can launch 300-yard drives on a regular basis, his status among fans is automatically elevated even when the rest of his game is suspect. John Daly became an instant fan favorite not so much for winning, but more for his ability to cream the ball. Some of this fascination with power is justified. It is the most effective way to demoralize an opponent; everybody, theoretically, can develop great touch, but not everyone can hit for tremendous distance. Superior putting, chipping, and iron play can all be cultivated with practice, but the power game largely relies on exploiting physical gifts. Furthermore, explosive power as it applies to golf is unique. The most powerful shotputters, discus throwers, and weight lifters can't pick up a driver for the first time and start launching 300-yard drives. They can generate more power than any other athlete, yet their type of power does not correspond to that exhibited in golf.

To understand the concept of golf power, it is first useful to understand what is meant by muscular power. Muscular power is the rate of energy expended, and it depends on the amount of energy available and the time taken to expend it. However, it can't be measured internally—that is, the rate at which muscles are actually producing and transferring mechanical energy—so the only real clues used to calculate muscular power are external: the amount of weight moved and the time involved. But even these external measurements are viewed differently by physicists and physiologists. Whereas physiologists associate power with mechanical force produced (Power = force X velocity), physicists define power as the rate at which work is being done or energy is transformed (Power = work done/period of time). Both define power as explosive force, yet time is the key variable for physicists. If a heavy weight is moved very quickly over a great distance, a lot of power is being generated. But if the same or an even heavier weight is moved more slowly over a shorter distance, then far less power is being generated. For physicists, Olympic lifters attain the upper limit of what is humanly possible. When they take 500 pounds from floor to overhead in 0.6 to 0.9 seconds, they generate superhuman power of 6,299 to 4,178 watts (8.4 to 5.6 horsepower).

Power for golfers is very different. Olympic lifters and professional golfers both impart the largest force possible for as long as possible, yet the muscle action is far different since the golf club weighs in at a few ounces and not hundreds of pounds. Olympic lifters try to move a very heavy weight a short distance quickly; golfers try to move a feather-light golf club a considerable distance at hypervelocities. Olympic lifters desire an explosive effort from almost all of the muscle groups simultaneously. Muscle power for golfers, on the other hand, is generated by the well-timed actions of select muscle groups. Golfers have the very difficult task of trying to keep the majority of the muscles out of the way, preventing them from interfering by contracting instead of relaxing.

The golf swing requires each body lever (bones) to come into action at the time that the prior one has reached maximum velocity. The muscle action involved in hitting the driver has much more in common with the fastball in baseball. In both, the larger, slower, and more powerful muscles of the lower body need to accelerate the smaller, faster muscles of the upper body and arms: the power comes from the ground up. A study of pitchers attributed 47 percent of pitch velocity to the step and body rotation, with the remainder coming from arm action, and the same probably holds for golf. According to calculations done by physicist Ted Jorgensen, author of *The Physics of Golf,* the typical professional golfer's swing delivers energy at about 2 horsepower, which requires 32 pounds of muscle or about one-eighth of horsepower per pound. Since muscles work in pairs, it requires 32 pounds of muscle mass and not just 16; for every muscle that produces motion in one direction, there is a muscle of about the same mass producing motion in the opposite direction. He reasoned that 32 pounds of muscle is too much to be generated exclusively by the arms and shoulders. Most of the power has to come from the legs, buttock, back, and abdomen. Unfortunately, most of us marginalize the contribution of the lower body because lower body movement is minimal and the speed of that movement is much slower than what takes place with the shoulders and arms.

Swing power is, in essence, fine-tuned control of explosiveness. That control includes making the most of muscle elasticity in the transition from the backswing to the downswing and developing a smooth transfer of momentum from one muscle group to the next. Neither are easy tasks to master, and that's why it is unlikely power athletes new to golf will ever master the myriad subtleties of the swing thoroughly enough to consistently hit the ball a long way. However, if the rules of golf were to change so that it was required that golf clubs weigh several pounds instead of only a few ounces, then these power athletes could quickly learn to outdrive the longest-driving pros.

Whips, Chains, and Momentum Gains

Merely examining the power of a golfer's swing is an interesting exercise, but not particularly useful in explaining the ultimate goal: maximizing momentum. In executing a powerful and technically sound swing if s crucial to maximize both the development of momentum during the swing as well as during the transfer of momentum from the body to the club and on to the ball at contact.

The dynamics that take place during a collision can be explained in terms of either energy or momentum. Momentum, which is defined as mass X velocity, determines an object's ability to keep moving with constant speed and direction, and is more useful in describing collisions because the combined momentum of the club and ball always must be conserved—momentum before the impact must equal the momentum afterward. This is not the case with kinetic energy (energy of motion) since some of the total kinetic energies of the collision may disappear into heat and sound.

Optimal momentum transfer comes from coordinating the

combination of linear and angular momentum. Linear momentum is generated by the acceleration of the body rocking forward toward the ball; angular momentum (rotation about a fixed axis) from the accelerating forces created by the rotational motion of the club and body. Although considerable momentum is developed, much of it is never imparted to the ball, and according to the conservation of momentum principle, all momentum developed has to go somewhere. Since the club does not stop the moment that the ball leaves the clubface, only a small fraction is transferred to the ball as the clubhead decelerates from roughly 100 mph to 90 mph, with the remaining momentum dampened within the body during the follow-through, as the club moves from the six o'clock position to directly overhead. There is not much that can be done to improve the actual momentum transfer at ball contact, yet there are a number of ways to minimize the momentum dampened by the uncoordinated actions of muscles during the downswing.

The timing of linear and angular motion to hit a golf ball is similar to the actions of a karate expert breaking boards or a boxer knocking out an opponent. A boxer tries to coordinate his angular momentum, turning with his linear momentum as he rocks toward his target. He wants to punch through the target, to concentrate on a fictitious point within the target, so that his punches terminate several inches inside the mark and not on its surface. To maximize the momentum transfer, a boxer applies the greatest force possible for as long as possible, as lengthening the period of contact makes his momentum more effective. In terms of the momentum equation, when a boxer punches through the body of his opponent, the mass of the blow becomes more important relative to its velocity. That's why a fighter who only flicks his jab and fails to get his mass behind the blow is never as effective.

The primary reason golfers do not effectively hit through the ball is the vastly different collision dynamics: In boxing, the principle is well demonstrated by a large area and mass (boxing glove with body mass behind it) hitting a much bigger area and mass (the opponent); while the comparatively small area and mass of the clubhead hitting a *very* small area and mass (the ball) does not make obvious the efficiency of this technique. The size and mass differences between the colliding objects account for why it is much easier to teach a fighter to develop the feel to punch through a target. For golfers trying to develop a similar feel for hitting through the ball, the difficulties are many. Primarily, unlike the boxer who feels his fist impact the target, the club is a "near feelingless" inanimate extension of the golfer. Secondly, the ball is a small and light target not at all resembling the boxer's large, heavy opponent; and whereas the boxer's blow goes from perhaps 60 mph to a stop, the clubhead slows only marginally, from approximately 100 mph to 90 mph. Finally, the collision lasts only 0.0005 of a second, which does not allow nearly as much time to feel the transfer of momentum. Because a boxer sees and feels the results of a good momentum transfer, it's easy for him to develop the motor memory and visualization skills to repeat that action. A golfer, on the other hand, is as likely to attribute a good drive to any number of things, such as a new wrist action, an exaggerated weight shift, or a change of ball position at address.

Physicists consider the golf swing a double pendulum, or two-lever action (see Figure 2.1). One lever, formed by the shoulders, arms, and wrists, rotates around an axis in the upper chest; the second lever is the club rotating around an axis formed by the hands. Coordinating and timing the actions of the two levers—the separate actions of the club, arms, and body—is what makes the swing's momentum transfer so devilishly difficult to learn. One-third of the way through the downswing, much of the body is accelerating, but by contact, that's no longer the case. Most of the body has slowed significantly because its momentum has been transferred to the arms and club; the arms are moving at a high speed, but the lower body—the link in the chain that's farthest from the action—has essentially come to rest. It's as if the body has

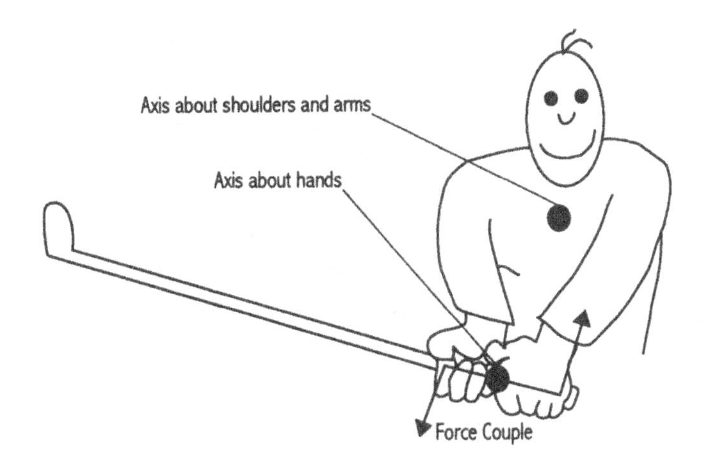

Figure 2.1:
The golf swing employs a double pendulum action: The shoulders and arms rotate about an axis in the upper chest, and the club rotates about an axis at the hands. Timing the seesaw action (or force couple) at the hands during both the backswing and downswing—the wrist cock and uncock—is critical in attaining a high-velocity swing.

hit a wall that does not allow it to rock forward any farther with linear momentum (because the weight is already out over the left foot) or twist with angular momentum (because the hips are already opened up). Instructional golf books and videos best describe this as "making sure that you get the body out ahead of the ball," which allows the ball to be on the clubface through more of the radius of the swing. Without getting ahead of the ball, it's still possible to make clean, square contact, but the momentum developed by the body will be diminished as it's transferred to the club and on to the ball.

In some ways, the two-lever golf swing most closely resembles the action of the nunchaku sticks, a weapon used in karate fashioned from two pieces of hardwood, each 10 to 12 inches long and held together by rope, leather, or a chain. To understand how connected rods transfer more momentum, a comparison can be made between nunchaku sticks and the typical billy club used by police. In combination, the nunchaku sticks are approximately the same length as the policeman's club, yet nunchaku develop far greater momentum (or impulse) transfer, which is why they are outlawed in many states.

Similarly, consider also the multirod bullwhip, in essence a collection of hundreds of very short rods, which makes it the ultimate momentum transferer. As each segment of the whip comes to a halt, it transfers all of its momentum to the following segment of the whip so that the snapping sound heard is the final segment breaking the sound barrier. Known as "cavitation," that cracking sound indicates the fast-moving tip is sending out a sound shock wave, no different than when a supersonic jet breaks the sound barrier.

Although a bullwhiplike momentum transfer is great in the abstract (if your body and club action were more like a bullwhip, you could count on adding at least 100 yards to your driving average), it's not easy to convert theoretical physics into physiological practice. Muscles and joints are flexible and capable of the coveted whiplike action, but bones are rigid levers, not multisegmented, and thus are far less cooperative. The best that can be hoped for is a chain action that harnesses the momentum of the multiple and various length levers. Optimally, one link or muscle group in this chain transfers its momentum to the next link, which then transfers it to the following link, and so on. When golf instructional books advocate the practice of "hitting into a firm left side" (assuming a right-handed golfer), they are describing this coveted process of momentum transfer. The firm left side is the end of the line, where each segment of the body slows or stops outright and takes the momentum it is given, adds to it, and transmits it along.

While the typical way to use a bullwhip is with an overhand motion, you can get a feel for the transfer of momentum by snapping a bullwhip or towel laterally with the left hand, which provides, in part, a sense of hitting into a firm left side. With the tip of the bullwhip or towel lying on the ground in a straight line behind you, move your arm forward to put the bullwhip in motion in a straight line. As the whip is pulled, the straight-line motion starts from the handle and proceeds toward the tip as the hand creates a force on the bullwhip, increasing its momentum. At a forward position, the hand stops while the rest of the bullwhip continues to extend along on its straight-line course. When the hand stops, the whip exerts a force on the hand that in turn decreases the whip's momentum, which is absorbed by the user. In a sequential fashion, additional segments of the whip come to rest. The crack of the whip occurs when all but the tip comes to rest, reaching the speed of sound as it snaps to a stop. At the start of the whip stroke, the total mass of the whip moves at a moderate velocity. But by the end of the stroke, a much smaller mass (just the tip) has absorbed the remaining momentum so that it moves at a much higher velocity. In other words, all the momentum generated by the body, arm, and hand imparted to the whip moves quickly to the very tip as each successive segment of the body and whip come to rest. The total momentum remains the same, but almost all of it has made its way toward the very tip.

A golfer's downswing works the same way. In the early part of the downswing, before reaching horizontal, the large muscles in the shoulder and back act in unison to begin releasing the coiled tension developed in the backswing. The positions of the shoulder, arms, hands, and club relative to one another remain unchanged early in the downswing as the hips drive forward, but once the swing begins moving horizontally, the relative positions change. Just like when working the bullwhip, some muscles are starting while others are stopping. And just like with the bullwhip, the action of the left arm is critical: How the left arm pulls the club along is just as important as the rotational acceleration (angular momentum) achieved relative to the shoulder joint. High-handicap golfers are prone to try to do too much with their right side, but in essence, the right side is overtaking or fighting against the left side, restricting the ability of the left side of the body to fluidly transfer momentum. The left arm should be "pulling" the body and golf club toward the ball; then as the left shoulder stops, the hip and legs more effectively "push" the body and club toward the ball. It is the combination of these two unnatural actions, a force-couple completed prior to contact that maximize the angular momentum in the downswing.

Teaching instructors rarely talk about momentum, instead they're always preaching swing velocity, which should be expected, since of the two variables in the momentum equation (mass X velocity), velocity is more important in determining the distance the golf ball travels. The significance of mass is diminished because the time of contact is too short to make it worthwhile to use a weighted clubhead; in fact, it's detrimental because the potential boost in distance from clubhead weight almost surely would be more than offset by much slower clubhead velocity. That's the case because the distance a golf ball travels is approximately proportional to the square of the initial velocity; in other words, when the initial velocity is doubled, the distance traveled increases approximately fourfold.

So, clubhead velocity at impact is key, and momentum transfer within the body is the means to that end: A well-timed transfer of momentum, from the ground up, achieves the highest possible clubhead velocity at impact.

• Heard It From the Pros

TIGER WOODS accurately explains the bodily momentum transfer this way: "explosive power starts from the ground up . . . flat-out, lower-body-initiated power . . . my legs and hips drive forward and my upper body simply unwinds" (*Golf Digest*, January 1999, p. 53). Although Woods deems this "explosive power," he really is describing the momentum transfer process. The lower body starts the whip action, the legs and hips driving forward deliver linear momentum, and the upper body unwinding delivers the angular momentum.

What About Acceleration?

It's a well-known fact that in order to drive the ball as far as possible, you have to hit the ball with as much velocity as possible. But, how important is constant acceleration (the rate of change in velocity)? At contact, is a still-accelerating 95 mph clubhead velocity superior to a constant or decelerating 95 mph? The answer here is unclear. One school of thought believes that an accelerating clubhead results in more contact through the contact zone, greater ball deformation (increased elasticity as the ball compresses and reassumes its shape), and less deceleration during contact so that more of the clubhead's momentum is transferred.

The 95 mph accelerating club will only slow to perhaps 90 mph, while the 95 mph nonaccelerating contact will slow more, as much as 10 mph or more.

The benefit of constant acceleration is more vivid if you consider the dynamics of the baseball collision. Mark McGwire, who in 1998 set the Major League record for home runs in a season with 70, does not have the fastest bat speed in Major League Baseball, yet it is highly likely his bat has the greatest velocity immediately after the ball is hit. At contact, McGwire's bat may be a bit slower than the quicker-handed Barry Bonds, who set a new single-season total of seventy-three home runs in 2001, but McGwire's constant acceleration and greater muscle mass result in less deceleration through the contact zone. Their results were similar, but the means to the end differed. Many sportscasters attest to McGwire's unique collision, a more resounding cracking sound, which results in more momentum being transferred to the ball. The crack of the bat comes from the combination of high-frequency bat vibrations and the outrushing of air from the ball as it deforms around the barrel, and in McGwire's case, both phenomena are attributable to greater momentum transfer and less deceleration through the contact zone than with the average hitter.

The other school of thought argues that the virtues of acceleration are being oversold. Since the collision time for a golf shot (0.0005 second) is so small a fraction of the total downswing time (lasting 0.15 to 0.20 second), the acceleration of the club on contact cannot change ball velocity very much. Moreover, acceleration indeed may be a significant determinate of how far a baseball travels, but the baseball analogy cannot be extended to golf since the collision dynamics are far different. Besides being larger, softer, and heavier than a golf ball, a baseball is struck by a bat traveling at 90 mph, while it's moving in the opposite direction at a similar speed, which results in a prolonged collision time.

Perhaps the acceleration debate is irrelevant anyway; if you can swing at over 100 mph, you are exhibiting the rhythm and timing to accelerate all the way through the downswing. If the clubhead doesn't accelerate all the way throughout the downswing, it is unlikely that the swing velocity is anywhere near 100 mph. A good golf swing, by definition, accelerates throughout the downswing. Instructors usually refer to this as good swing tempo—a smooth constant acceleration.

The major culprit preventing golfers from achieving smooth, constant acceleration is usually poor mechanics. Kinesiologists, scientists who study how the body moves, call the golf swing a kinetic chain event; that is, the muscle groups in different parts of the body work in series, one after the other. And to get the most out of this series—create the greatest momentum to drive the ball—if s essential for one muscle group to activate at the moment just after the previous one applies maximum velocity to the club. Most recreational golfers, however, reach a top speed too soon, and as a result, the clubhead actually starts to slow before contact. Rhythm and timing problems of this nature are not unique to golf; for example, few people realize that world-class sprinters usually clock their fastest speeds in the 100-meter dash between the 50- and 60-meter marks and almost always record a slower time between 90 and 100 meters (because acceleration takes place over the first 50 meters, while over the last 50 meters the objective is to maintain the top speed). Both sprinters and golfers require incredible precision timing so that the hundreds of muscles involved turn on and off in the proper sequence. And the deactivation is as critical as the activation; otherwise muscles contract or "fight" against each other, retarding acceleration. It's common to attribute a herky-jerky swing to a lack of flexibility when in reality if s more often a problem of misfiring muscles (that is, the muscles contract too early or too late).

Good rhythm and timing are nothing more than getting muscles to fire on and off on cue and in the proper sequence. When they don't, your swing will always have a herky-jerky element to it. And when the chain action is out of sequence, it prevents acceleration all the way through the contact zone, which makes it impossible to achieve a high velocity at impact.

Why Not Overswing?

Although greater velocity and constant acceleration are extremely desirable, more is not better when it comes at the expense of accuracy.

Trying to kill the ball heightens the probability of an errant shot for several reasons: First, when you try to blast the ball, the greater muscular effort is there, but it is not all going toward good ends. Blasting usually results in rushing, overstretching of muscles, timing difficulties (between rotational and linear efforts as well as muscle actions), and balance problems. Secondly, calling on every muscle in the body to generate as much force as possible significantly lowers the probability of striking the sweet spot—the maximum power spot, the zone with the highest elastic energy. Hitting the sweet spot is critical, and when it's missed, much of the energy intended for the ball is instead diverted to club rotation, and the twisting force for any given mishit is going to increase with increasing swing velocity. Furthermore, if you normally are around the edges of a club's sweet spot, blasting increases both the probability of a mishit and the severity of the error.

Of course, big swinging pros and hackers alike feel emboldened by the latest oversized drivers since they improve the margin of error for impacts that are slightly off center and are far more forgiving of contact at the fringes of the sweet spot. Though these clubs decrease the perils of blasting, the risk still far exceeds the reward; instead, it's better to concentrate more on swing tempo and constant acceleration. Improving the rhythm and timing of your swing will give you consistently better results.

The Where and When of Weight Shifts

Effectively and properly transferring weight is essential for maintaining good balance, coiling naturally, and transferring maximum momentum to the ball. Conversely, shifting weight improperly usually results in detrimental countering actions such as dipping at the shoulders, under- or overrotating, and leaning too far forward or backward from the hips.

Surprisingly, the weight shift that actually occurs is not what most golfers think. During the backswing, many golfers are under the mistaken belief that the major weight shift should be directly sideways—from the left foot to the right foot—during the backswing. But what actually should be happening is a shift from toe-to-heel for the right foot and, to a lesser degree, from heel-to-toe for the left foot; there is very little, if any, backward and forward motion. This shifting takes place so that our center of gravity (a major determinate of balance located slightly above the waist) remains stationary as the hips turn, countering the shift caused by bringing the arms back during the backswing.

During the downswing, weight is redistributed from toe-to-heel of the left foot, and to a lesser extent, from heel-to-toe of the right, which is necessary for rotation of the hips, back, and shoulder. By the end of the downswing, the majority of weight is distributed over the left heel, and without this weight shift, much of the momentum developed in the powerful leg and body muscles cannot be transferred on effectively to the quicker and smaller muscles in the arms. Usually if the weight distribution is correct, the swing will be well-balanced, fluid and should result in a higher velocity impact.

Physicist Ted Jorgensen in *The Physics of Golf* calculated the clubhead-ball impact speed for swings involving a well-timed weight shift to be 14.4 percent greater—170.5 feet per second (52 meters per second) with the proper shift and 149.0 per second (45 meters per second) without it. Given the same stroke motion, then, and assuming that the complex two-lever motion of the swing is well-coordinated with the weight shift, a drive that would travel 200 yards (183 meters) without a weight shift will travel 229 yards (209 meters) with one.

Historically the weight shift toward the front foot has been overrated, especially when emphasized over the heel and toe weight shifting necessary for rotation. It is impossible to rotate effectively and transfer momentum from the lower body to the shoulders and arms by solely rocking back during the backswing and rocking forward during the downswing. To achieve the body rotation that will not impede the momentum transfer process, heel-to-toe and toe-to-heel weight distribution must also take place.

Spring-loading: The Backswing and Transition

Elastic, or spring, energy is an important aspect of many sports, and golf is no different. In some sports the generation of elastic energy is exclusively the domain of the equipment. Pole-vaulting is one example: A pole-vaulter sprinting toward the bar has kinetic energy, which is stored temporarily as elastic potential energy, or the energy of deformation, as the pole bends. As the pole straightens, releasing this energy, it helps launch the pole-vaulter over the bar. The greater the bending, the greater the elastic potential energy; thus, elastic potential energy (from increased pole bending) is greater for heavier pole-vaulters and those who reach greater speeds during the run up.

For sports like tennis and golf, elastic energy comes from the body as well as the equipment. The preparatory body and equipment motion (such as bringing the racket back and overhead to serve in tennis and the backswing in golf) is used to stretch the muscles, which are much like rubber bands in that the greater the stretching, the greater the distance and quickness of the muscle contraction. A properly executed backswing "spring-loads" the downswing, making it vastly more effective. You can hit the ball a great distance by starting with the club stationary overhead (just as a diver can spring from the board from a standstill), but nowhere near as far as with a properly executed backswing. In reality, the backswing is far more important than the downswing, because a properly executed backswing determines the wrist cock, the weight shift, and the rotational action. If everything goes right with the backswing, a properly executed downswing occurs almost naturally.

If you have time to practice only one thing, it should be the development of a smooth rhythmic transition from the backswing to the downswing. Loading occurs during the last instant of the backswing, the split second when the body starts going forward as the clubhead is still on its path back. A properly executed transition is like the difference between bench-pressing from an at-rest position compared to lowering the weight quickly and getting an elastic bounce or rebound off the chest so that the inertia of the barbell assists the lift. Thus, an expanded definition of good swing tempo is: a backswing-to-downswing transition that stores the greatest elastic energy so that acceleration is quick and consistent all the way to the ball.

The backswing should be initiated and controlled by the left arm and elbow so that the body more naturally rotates about the base of the spine. Bringing back only the arms and hands, loads up only those fast-moving muscles and not the larger more powerful slower-moving muscles of the lower body. Correctly loading up on the backswing creates a rotational power source as the hips and shoulder rotate. Your muscles are fully loaded when you feel coiled tension, or stretching, throughout much of the left side of your body as weight is shifted to the right heel.

Cocking and Uncocking

Millions watch in awe as Major League All-Stars like Gary Sheffield and Sammy Sosa use a pronounced wrist cock and a whippy uncocking action to blast mammoth home runs deep into the bleachers. Since many do not start playing golf until later in life, it's only natural for them to fall back on the Little League Baseball swing of youth. This background, coupled with the adoration of Major League power hitters, can lead to the development of some unfortunate misconceptions about how to hit on the course, the worst of which is the mistaken belief that cocking and violently releasing the wrists is a sure way to crush a golf ball.

Of course, driving a golf ball and hitting a baseball do have much in common: Both require a two-handed grip, a raising of the hands, a twisting of the body and involve the collision between a club and a ball. But that's where the similarities end. Because the golf club is much lighter and longer than a bat, the contact is at the end of the shaft and not in its center, and the golf ball is a stationary target and much lighter than a baseball, the dynamics are very different.

Although wrist action is a complex, multifaceted phenomenon, there are two certainties: The cocking is more important than the uncocking, and it's unwise to try to uncock your wrists like a baseball hitter. The standard instruction is to cock your wrist before starting the downswing. While some instructional books claim it doesn't matter whether you consciously cock during the backswing, in reality, it does matter. The cocking should occur naturally because this action best helps generate and store elastic energy for release during the downswing.

Slow-motion replays show that the hips of professional golfers actually start moving forward before the club comes to a rest at the top of the backswing. Physicist Alastair Cochran, author of *Search for the Perfect Swing,* determined that the forward motion of the hips should begin approximately 0.1 second before the clubhead reaches the top of the backswing to store the greatest "coiling power." In short, this action and counteraction—the body coming forward as the club is still moving backward—dramatically improves your ability to store and then release elastic energy. The body and club moving in opposite directions is subtle, counterintuitive, and hard to visualize; an easier visualization is to bring back and then whip forward a towel with a left-handed underhand motion. You can feel how much easier it is to snap the towel by starting forward with the hips and shoulder well before the towel has gone all the way back. If the wrists are cocked prematurely, a considerable amount of elastic energy never can be generated.

For a moment, let's compare the muscle elasticity dynamics that occur for a drop jump (from a chair to the ground) rather than a stationary vertical jump. Higher jumps will always result from a short drop jump since the muscles will be stretched farther from the bounce: The velocity and mass of the body stretch the leg muscles on contact so that they can contract over a greater distance for the jump. The wrist cock works similarly. Jorgensen studied swings with wrist cock angles of 90, 110, and 130 degrees and found that the larger the wrist cock angle just prior to the downswing, the greater the swing velocity. Of course, this assumes good flexibility so that the wrist cocking is natural and comfortable. If the exaggerated wrist cock angle is not natural, the elastic energy will not be stored properly and it will be very difficult to maintain a good swing tempo.

If a camera were positioned above a golfer, the backswing would appear as if he were coiling himself like a spring around an invisible vertical axis. To assist in this twisting, the left arm remains straight while the right arm folds toward the body with the elbow pointing down and back, not out and up. Teaching professionals recommend a pushing action led by the left elbow so that the body fluidly rotates about the base of the spine. This subtle motion creates a rotational effect around the vertical axis that leads the wrists to cock automatically until the backswing is finished; how readily the wrists cock is mainly determined by wrist and backswing flexibility.

Aside from mastering the backswing skill of cocking, there is the completely unrelated question of downswing uncocking—pushing down with the back wrist and releasing it prior to contact. As velocity increases during the downswing, there's a natural tendency to uncock early due to the increasing pull or torque at the wrists. At the beginning of the downswing, there is only 15 to 20 pounds of pull, or torque, from the rotating club, but nearing ball contact, the club, now moving at around 80 to 100 mph, quickly increases the torque to anywhere from 70 to 90 pounds. This fourfold torque increase makes you feel like letting your wrists go, but don't. This is one instance where you should definitely resist what feels irresistible and natural. Moreover, delaying seems counterintuitive since uncocking in most sports results in an extra power boost; skillful uncockers rocket home runs, hit the biggest serves in tennis, and blaze overhead smashes in badminton at speeds exceeding 200 mph. However, the longer uncocking is delayed, the better. Toward the bottom of the downswing, the larger more powerful muscles of the lower body actually slow so that the left arm and club can catch up, which improves the transfer of momentum from the club to the ball. Any golfer who uncocks early hampers this final action in the multilevered swing, dampening angular momentum within the body. An early uncocking would be like the last four inches of a bullwhip advancing before the middle section has had a chance to come forward. When this occurs, momentum is lost, and the whip cannot snap.

There is no truth behind the myth that there's an imperceptible power-boosting uncocking going on for big hitters. It's nearly impossible to time the uncocking for an additional power boost because a golfer's arms act through a very large angle before ball contact occurs. The long arc of the swing makes it more effective to have the hands well in front of the ball at the point of contact rather than poised for last second uncocking.

• Rubberband Man

IN the hope of achieving a better "whip action," some golfers try using a highly flexible shaft and loosening one hand at the peak of the backswing for a more rubbery feel. This is a dubious approach because it usually shortens the swing radius. When you loosen the grip of one of your hands, it results in asymmetric pressure on the club, which tends to uncock the wrists early. An early uncocking also means that the clubhead is too far ahead of the body at the time of contact, which leads to a less effective transfer of momentum.

A Soft Grip Before YouRip

A golf club is not a sledgehammer, so it shouldn't be gripped that way. Whereas a sledgehammer usually weighs ten pounds or more, a golf club usually weighs about a pound. Without a tight grip on the hammer, it's going to fly out of your hands; meanwhile, the lightweight golf club allows for everything from a death grip to a feathery touch. At either extreme, you won't lose your hold on the club.

The primary physiological reason against the death grip is that it tends to shorten muscle contractions. To demonstrate this for yourself, grip a club as tightly as you can; you should feel tautness in the muscles in your forearm and even as far up as your triceps. Taut arm muscles during the backswing don't bode well for the generation of swing velocity in the downswing because a good backswing relies on efficiently storing elastic energy in the muscles and tendons as they stretch just prior to the downswing.

According to the law of conservation of energy, the energy of the club as it works its way up and overhead, can neither be created nor destroyed, but it can be transferred. While some is dissipated within the body and some is lost as heat, the rest is stored and returned as spring energy: If bones, muscles, joints, ligaments, and tendons are considered in combination, they have some of the same mechanical properties as a stiff spring, as well as the dampening properties of a sponge. So, the more you can take advantage of the springlike properties of your muscles, the more efficient and effective your swing execution. A very firm grip, which necessitates taut muscles, results in more dampening and less elastic energy being transferred between the backswing to the downswing. Further, if you do not allow your muscles to fully relax (stretching out to their greatest length), the distance and speed of contraction is limited. It isn't surprising that studies of grip strength have found that amateurs hold their clubs much more tightly than professionals; during the backswing, pros exert about 25 percent of the maximum force, whereas amateurs' grips are much firmer, up to three times as tight. The same trend holds for the downswing, but the difference is not as dramatic.

Besides hampering the ability to make contact at the greatest velocity possible, professionals avoid the death grip because it also compromises control and accuracy. A relaxed grip is crucial for touch shots, because regardless of the sport, touch shots require an activation of motor control memory. That's why basketball players bounce and rotate the ball in their hands before shooting a free throw; it's not just to channel some of their nervous energy. Motor memory is sharper after rotating and bouncing the ball because that practice activates the fingertips' nerve endings. The same technique is employed to great effect by golfers: Rolling the hands around the grip before settling on a final hold helps activate motor memory, and by maintaining a relaxed grip, the communication between the fingertips and brain will be more effective, and the result will more closely mirror preswing visualization.

Some scientists, though, are skeptical of the light grip theory: Since the centrifugal effect (the flinging of the club outward from the hands) can exceed 100 pounds for faster swingers (only 30 to 40 pounds for slower ones), they feel a light grip is untenable because a tighter hold is needed to counteract the centrifugal effect of the clubhead. Because the 100 pounds of force pulling at the hands occurs near the contact zone, and it's only a major factor for shots hit with the longer-shafted driver, it's probably more accurate to describe the ideal grip as being as light as possible, and not uniformly feathery light all the time. From a physiological perspective, only a grip that is as firm as it has to be will generate the greatest clubhead velocity.

It should be noted, however, that there are a couple of caveats about maintaining a relaxed grip. If you suspect a high probability of missing the sweet spot or are powering through heavy rough, then a death grip near the contact zone obviously makes sense. Of course, lack of confidence is a factor too. If every time golfers step up to address the ball they think about mishitting, the instinct is to always grip the club more tightly.

• The Grip Strength of the Pros

SAM SNEAD, in *The Game I Love*, said "You could probably take my club away from me at the top of the backswing, I gripped it so lightly. I think with light grip pressure, you can get more zip and better release at impact for more distance." True. The lighter your grip at the top of the backswing, the greater the muscles in the arms stretch and the farther and faster they can contract; thus, you do have more zip in your swing at impact.

Hand position is another aspect of the grip that is very important because it determines and controls the axis at the second lever of the two-lever swing (again see Figure 2.1). You need a grip that is both comfortable and efficient to hit the ball with great rhythm and tempo; therefore, any type of overlapping grip, as opposed to a baseball grip, is preferable from a physics standpoint because it shortens the lever at the hands—it reduces the distance from the little finger on the left hand to the index finger on the right hand. The hands work better together, which benefits both power and control.

Why the Backswing and Downswing Path Shouldn't Match

Some instructors will tell you to try to follow the same path for the downswing as you used for the backswing. But try as you may, this is almost impossible. Moreover, for golfers with a slice, chances are it's doing more harm than good: In fact, the conscious effort of trying to match the downswing path with that of the backswing just taken might be the root cause of the slice! Though you may think that you're following the same path, that's in reality very unlikely. When the clubhead is at about the five o'clock position (six o'clock is resting against the ball), it's moving at about 5 mph during the backswing and at 80 to 100 mph as it returns during the downswing, which is a huge difference. And although the club only weighs about a pound, the angular momentum of the club—which depends upon its velocity, mass, and the distance and distribution of the mass from the axis of rotation—will be 16 times greater during the downswing. Maintaining the same path requires a mental adjustment to account for the feel of different angular momentum. The hammer throw is an extreme example of the same principle because the hammer weighs 16 pounds and is located 4 to 5 feet from the axis of rotation. Even if a hammer thrower tried, he could never maintain the same plane for the hammer on his second and third revolution as he did for the first revolution, since the hammer's momentum is as much as 60 times greater on the third revolution as it was on the first.

Because of the much greater clubhead momentum on the downswing, you should forget about following the same path and instead let the weight of the club (its momentum) set your hands underneath the shaft. Instructors describe this as an early "powering of the club inwardly," that is, swinging closer to the body. This action should compensate for the downswing's greater angular momentum.

As mentioned, the root cause of a horrific slice is likely the result of trying to follow the same path. A slice is caused by sidespin—the clubhead moving in an outward-to-inward path as it meets the ball. As the angular momentum of the club pulls the club in a more outward plane during the downswing, eye-hand coordination will make the adjustment so that the clubhead strikes the ball, yet to do so, the clubhead has to take on a serpentine motion late in the swing to come back and hit the ball. Most golfers have no idea that this slithering swing path creates a slice because they mistakenly think the club is following the same path on the way down as it did on the way back, but the vast differences in angular momentum make it impossible to match the two trajectories. You must compensate, and the earlier the better. What instructors refer to as an early "powering the club inwardly" motion to avoid the late eye-hand coordination adjustment is the best approach, and the higher the swing velocity, the more important the adjustment, because the greater the angular momentum, the more difficult it is to change the clubhead's speed or direction.

Another major cause of sidespin is the orientation of the clubface to the ball. If you are making a conscious effort to power the club inward and still find yourself slicing, your problem may be with your grip. Without changing how you grip, try to make tiny adjustments where you grip by rotating the club in your hands until you find the orientation that gives you the most consistently straight shots—the clubface is repositioned relative to the ball, but the orientation of your hands and stance remain the same. After discovering the grip and orientation that result in sidespinless shots, then make the tiny adjustments in your stance so that the ball follows a straight line path down the middle of the fairway.

Eliminating your serpentine swing motion and adjusting the clubface orientation to the ball at address are the best ways to put an end to your slicing days.

The Swing Plane and Length

Among the experts there has been much writing but little agreement about the correct swing plane (angle of shaft) and backswing length (how far back to go). The truth is that there is no one correct swing plane; even among professionals, the angle of the club shaft varies considerably. Some pros use an upright, vertical swing (involving a more underarm, rather than sidearm motion); others use a flatter, more level swing. However, despite the clubhead path varying tremendously, the path of the hands is consistent. This, in effect, illustrates that all players use similar movements of the body and arms to generate the greatest momentum possible, and the swing plane chosen has no effect on the ability to maximize the momentum transfer. In fact, the actions and timing of the legs, body, and arms, which are critical in determining the clubhead acceleration and final velocity, can be identical for the two swing plane extremes.

Much of the difference between swing planes is natural, attributable to physiological and mechanical differences in the movement of the body's levers. The foremost determinate of swing plane is the relative rotation of the hips and of the shoulders. Hip rotation alone results in a very flat swing; on the other hand, when shoulders are emphasized alone with little hip rotation, the swing will be very upright. Neither is intrinsically superior, yet the very flat or very upright swing seem to indicate that clubhead acceleration is being retarded by the shoulders or hips, respectively.

Although neither style is superior for generating greater clubhead velocity, a more underarm swing would be helpful for players trying to correct a slice. The flatter the swing, the greater the angular momentum pulling the clubhead farther from the intended path, which then would require a more pronounced downswing correction to strike the ball as squarely as possible. The more upright swing has equal angular momentum, but more of the momentum is in the vertical plane than the horizontal, and thus, it should be easier to bring the clubhead to the ball in a straight line and square at contact, minimizing sidespin.

Unless you're struggling to correct a slice or hook, there is no need to change your swing plane. Everybody comes in different sizes, shapes, and flexibilities, so the most prudent advice here is to stick with what works best for you. If you have wide shoulders and narrow hips, it is more natural to have a flatter swing; conversely, if you are smaller at the shoulders and wider at the hips, it is probably best to be more upright. For those considering flattening out their swing for the sole purpose of using an extra-long shafted driver to generate more angular momentum, forget about it. For a discus thrower who rotates two to three times besfore release, a discus held 3 feet from the axis of rotation in the body rather than 2.5 feet will generate 20 percent more velocity. However, for the golfer, the longer the shaft, the greater the force needed to get the club in motion. Lengthening the shaft has the potential to create greater angular momentum like the discus thrower, but it probably only marginally adds to the velocity (1 or 2 percent). That's because the arc of the swing is too short to benefit much from a flatter swing and a longer shaft. In fact, without also lightening the clubhead, weaker and less quick-handed golfers are likely to generate less velocity.

Swing length—how far back to bring the club—is another aspect that golfers at all levels regularly consider changing. Intuitively, golfers feel that the higher and farther back the backswing, the farther the ball will travel. In theory, the longer the swing, the greater the force exerted on the club. The few degrees of extra arc that the exaggerated swinger uses allows for a few extra fractions of a second to exert a force on the club. Though the potential for greater distance is definitely there, studies have found that the extra time and distance of the backswing results in only a very marginal increase in clubhead velocity at impact. The reasons for this are simple: First, greater length has the potential to mess up your timing, which is particularly relevant if you play less than two or three times a week. Secondly, an exaggerated backswing has the potential to overstretch the coiling action of the body ruining your swing tempo; the fluidity to smoothly accelerate through the downswing is badly compromised. And finally, the longer clubhead path may require an adjustment in the downswing to ensure an accurate collision with the ball, and midswing adjustments make it impossible to maintain a smooth acceleration. In all, the risks to directional accuracy far outweigh the marginal increase in drive distance. Given the choice, you should pay far greater attention to what you are doing with your wrists rather than how far back your club goes during the backswing. Many young professionals use an exaggerated backswing, but as they gain experience and take on a swing coach, they usually shorten it up a bit, discovering that for a slight sacrifice in length, their drives will find the fairways far more consistently. Obviously the shorter backswing is even more beneficial for recreational golfers, and more importantly, it is probably the simplest instruction to put into practice.

• How Long A Backswing?

In *Golf My Way*, Jack Nicklaus claims you have to go back as far as you can and challenges anybody to name a great champion who ever won consistently without a big backswing. Jack's way probably should not be your way. It takes a lot of practice to squarely strike the ball, and the longer the backswing, the less likely your chances of striking the ball squarely. Most golfers are better off shortening the backswing, and instead concentrating on developing good rhythm and timing with a natural wrist cock at the top of the backswing.

3

Mind over Muscle:

Motor Control and Mastering the Mental Game

WHEN commentators and swing doctors talk about golf being more mental than physical, what they are really saying is that the brain's control over the actions of the joints, muscles, and tendons has more to do with success than any other factor. Pure physical talent may endow you with the potential to drive a ball 300 yards straight down the middle of the fairway, but if your mental approach to the game is questionable, then you probably will only achieve this result sporadically. On the other hand, someone who has less God-given talent (possessing only the ability to drive the ball 280 yards), may have a superior mental game that results in far greater driving consistency.

The mental game—how the mind affects the body's performance—has always been a bit of a mystery; primarily it revolves around the nervous system, which coordinates complex movement and directs the muscles. This motor control is often studied in the context of speed or precision, and golf is a sport that requires both: When a golfer sends a bullet-drive 300 yards down the middle of the fairway, it is an example of both speed and precision (speed enough to generate clubhead velocity in excess of 115 miles per hour and the precision to squarely strike the ball so that it carries in the intended direction). Some strokes require primarily speed (the drive), others primarily precision (the putt or chip), and many a more equal combination of both (a 6-iron fade around a dogleg). However, motor control skills are most noticeable in the short game, since these shots call for fine-tuned accuracy in order to hit a very specific target rather than an explosive swing directed less meticulously. A drive from a tee does not entail as many variables to analyze, and the shot is focused on a much bigger target.

Before delving into different aspects of motor control, it is useful to first briefly consider the basic workings of the nervous system that are involved. The nervous system is the general term for the network of cells extending from the brain throughout the body that communicates information in the form of electric impulses that travel back and forth between the brain and the body's periphery. Our brains are "headquarters"—the command and control processing center—for all movement, from the very simple (walking) to the very complex (dancing or playing a musical instrument). Besides directing activity, the brain also works simultaneously in reverse as the main switchboard, gathering internal and external signals, which includes constant feedback from the sensory system: visual signals, touch signals (primarily from the fingertips, but also sensory receptors in the joints and muscles), auditory indicators, and balance and orientation calibrations from the inner ear (vestibular system). Our brains analyze and process visual information and then integrate it with information from the other senses, to "map" out quick and efficient performance instructions for our bodies. Continual feedback goes on between the senses and the brain for correction, adjustment, and refinement. Although that communication has only a marginal effect on the shot being executed (there is not enough time to make adjustments such as changing grip pressure in midswing), it's a vital means of improving future performance by monitoring what's happening during the stroke. The sensory receptors in the joints and muscles store this feedback information for replication in the future, and since the golf swing entails thousands of subtle movements, there are thousands of messages sent about the position of the joints and the contraction and relaxation of the muscles. Therefore, developing a good mental approach involves a two-pronged approach: fine-tuning the messages being sent out, and learning how to make the most of the messages being sent in. Awareness of both aspects is vital if you are to gain confidence as you gain skill.

Any time spent working on the mental game is time well spent, because the mental drives the physical, and the mental game is about getting the brain and body on the same page. Developing the right mental approach is a far more constructive way to save strokes than any other, including investing in new balls, clubs, or even a plethora of lessons from a swing doctor.

Preswing Visual Analysis

You never want to square up to swing until your eyes figure out where you want the ball to go and you believe that the club in your hands will send it there. You want to start by taking as many mental notes as possible; the more careful and complete the analysis, the better the chances of success. With every shot, go through all the variables: lie, wind conditions, length and dampness of fairway, condition of the rough and green, placement of the flag on the green, and how you're hitting the ball that day. You should also keep in mind that shots demanding great precision require more comprehensive analysis; a putt from 15 feet necessitates far more scrutiny than does a drive from the tee. It's equally important to remember that no two shots are the same; you may play the same course hundreds of times a year, but conditions always change. You may have had similar shots under similar situations, yet none of them is identical.

As in most sports, sight is the most important sense in golf. This is a reflection of fundamental physiology since eyes provide the majority of sensory information reaching the brain, and the primary visual cortex of the brain is up to five times larger than the area related to all the other senses combined. Our eyes serve a reactive role in telling us what they see and an active role when told what to look for. Although visual processing skills are largely hereditary, research has demonstrated a strong learning component that has led to the development of training programs teaching athletes to better recognize what they see and what they should look for. From participation in sports, the eyes and the primary visual area of the brain naturally become more highly developed, but vision training can lead to even better performance. Similar to strength training, these programs isolate and work on individual visual skills, improving them beyond what can be achieved simply by participating in sports.

The major goal of the preswing analysis is to provide the most complete and accurate visual imagery to more fully activate the brain's visual cortex. Out of over 20 visual skills used in sports, the most important for preswing analysis are static acuity, spatial localization, and fixation. Static acuity, the most widely known trait, refers to the clarity or sharpness of a stationary visual image and the ability to distinguish separate parts of a target. Since neither the ball nor the golfer move during the swing, dynamic acuity (which would be required if either the object or the athlete were moving) is not a concern. Whereas significant improvements in dynamic acuity can be achieved with training, static acuity is primarily determined by heredity. But thanks to modern medicine and technology, excellent static acuity is readily available as long as we have regular vision checkups and wear corrective lenses when necessary. As a general rule, if static acuity drops below 20/40 (20/30 is considered average, 20/20 good, and 20/10 excellent), sports vision experts recommend a prescription for corrective lenses.

An alternative to corrective lenses that is becoming increasingly popular among notable athletes, including golfers, is laser assisted in-situ keratomileusis (LASIK) surgery. This procedure reshapes the patient's cornea and gives over 80 percent of them 20/20 vision. However, LASIK surgery is not risk-free: It can result in pain, haziness, poorer night vision, slower glare recovery, and an increased risk of globular rupture (an explosion of the eyeball) because surgery weakens the eye. Since night vision and glare recovery speed are vision traits that are inconsequential to golf, and it is not a high-risk sport in terms of eye injuries, the dangers associated with LASIK surgery are much less than for athletes in other sports. Nevertheless, it's a bit surprising that Tiger Woods, who is destined to earn tens of millions of dollars a year from tournament winnings and endorsements, would opt for LASIK surgery at this time. The procedure only started to be performed in the early 1990s, so there have been no studies tracking side effects beyond a few years. It may prove to be relatively safe and effective in the long-term, but the risk for a golfer like Tiger Woods does not seem justified when contacts can provide the same or better static acuity with no long-term risks. Perhaps for Woods it was nothing more than a confidence booster; however, to date, there's no conclusive evidence that LASIK-corrected vision results in superior sports performance.

After trusting your eyes to analyze the upcoming stroke, you must then fixate, or focus, on the target. Your brain and nervous system's response is best when your eyes take a "hard," fine-centered focus on a small, precise target. But until you are ready to begin the backswing, stay with a soft focus to avoid staring at the ball too long before the stroke begins; overfixating, or staring, diminishes eye-hand coordination abilities. Putting is a little bit different: Because they almost always break, when analyzing putts you never want to focus too much on the ball or the cup, unless it's a very short putt. Instead, it's best to fixate on a midpoint target—usually an area of the green that is 10 to 15 feet from you.

Spatial localization, which refers to the ability to judge where you are relative to other objects, is another important visual skill. It involves more than just the ball's lie and where you want it to go, but also the location of hazards along the intended route. Poor spatial localization is largely a problem of perception. The perceived world differs from reality. Golfers who tend to regularly play short of the intended target suffer from esophoria, while golfers who regularly play long may suffer from exophoria. The hole appears, respectively, shorter and longer than it actually is. Likewise, some golfers perceive the hole to be either to the right or left of its actual location. (Think about bad sunglasses: If you quickly flip the lenses up on your brow, you see how they make objects appear closer than they actually are. The lenses of some people's eyes do the same thing.) Most golfers with poor spatial localization do not even realize they have a problem; instead of blaming errors on misperception, they are more likely to point to poor execution or the wrong club selection as the culprit. If you suspect a problem, you should see a sports optometrist who not only can diagnose the problem, but also design a training program to improve your spatial localization.

Depth perception is closely related to spatial localization. Although vital in determining the yards from tee to green and distances to the hole when putting, yardage books, course markers, and caddies who know the course all minimize the need for acute depth perception. Other visual skills—dynamic acuity, eye tracking, eye-hand coordination, peripheral field awareness and accommodation (that is, the speed of focusing) as it relates to visual reaction time—are far less of a concern for golfers. That's because in golf, as opposed to baseball and tennis, you have the luxury of taking as much time as you like to strike a stationary ball.

• The Pros and Cons of Sunglasses

EXPECTING to spend long hours in the sun, more recreational and professional golfers than ever before are wearing sunglasses. Besides the eyestrain and fatigue that come from sun glare, excessive exposure to 290 to 400 nanometers ultraviolet light (UV-A) and 400 to 510 nanometers blue light (UV-B) can cause cataracts.

Sunglasses are most beneficial in locations where the sun's rays are most intense, such as at high altitudes or areas where sunlight reflects off water and sand. But since golf courses are primarily green, absorbing much of the sun's rays, sunglasses are not critical for golfers. That's why most PGA golfers are very reluctant to wear them, fearing that they may be detrimental to performance by altering depth perception, limiting peripheral vision, and obscuring shadows. Depth perception is a problem with all shots (for example, the lenses will make the ball look to be at a distance of 62 inches or 58 inches from the hole instead of the 60 inches that it is), while peripheral vision and shadows are chiefly putting concerns. In particular, the better your peripheral vision, the easier it is to map out a course for long putts, and seeing shadows is very helpful when trying to read greens, because the grains and slopes of greens change shade when viewed from various angles.

For frequent recreational golfers, sunglasses are strongly recommended. There are models available that significantly reduce sun glare, provide measurable ultraviolet ray protection, and do not obscure shadows. Look for compliance with the American National Standard Institute (ANSI) standards for sunglasses (labeled ANSI Z-80.3).

Once you decide to try sunglasses, it's important to give yourself plenty of time to become comfortable so that you are confident wearing them. The reason David Duval is very comfortable wearing sunglasses is because he began wearing them in his youth. For the same reason, the majority of pros won't wear sunglasses because they never have; the feeling is, if it ain't broke, don't fix it. At minimum, however, you should wear a cap to shade your eyes from as much direct ultraviolet and blue light as possible.

Visualization: Picture the Process

Visualization is probably the most underappreciated visual skill used in golf, but it shouldn't be. You can't be a good golfer unless you're a confident one, and confidence comes from conviction in your visualization. The ability to recall similar shots made under similar circumstances, picture the entire event, and then execute, is absolutely vital.

Visualization can be described as the memory's search for information, followed by the application of that information. In golf the process involves the search for swings made under comparable conditions and then picturing the full sequence of the event. In general, the imagined shot starts with contact, followed by flight, then landing, and finally, the roll to a final resting spot.

For high-precision putting, a far more extensive visualization process is warranted:

1. **Develop a visual picture of the desired action, from start to finish.** Visualize the contact, the follow-through, the breaking roll, the approach to the high side of the cup, and then the ball dropping into the cup. You should always visualize the approach from the high side. It expands the size of your target, which is especially true for big breakers, which present the greatest opportunity to slip in the side door.

2. Never **take your** eyes **off the ball on its journey toward the hole.** Concentrate on remembering everything that takes place, it's an important part of the process for playback. The less you

watch, the less you concentrate, the less you internalize for future playback. Because of brain clutter and a lack of concentration, the majority of people don't do a good job of remembering: Usually when someone claims that they forgot something, what really happened is that they didn't make an effort to remember it in the first place. Everyone uses visualization—it's one of the primary methods of learning and remembering—but because it's a subconscious process, they're usually unaware of it. Thus, the best way to improve visualization is to practice in such a way that the thorough imagining of a shot before hitting the ball is a conscious effort—you want to remember more of the subtle details of your every action.

In trying to visualize a putting stroke and developing a visual memory, accommodation, or speed of focusing, is a key skill. When picturing the putt, you have to shift focus from the ball to the hole, or from the ball to several intermediate targets along the way, and finally on to the hole. The quicker you can shift focus along the path of the intended putt, the more confidence you will have in your visualization.

Speed of focusing can develop into a problem for older golfers using bifocal lenses since the reading lenses are designed to focus at 13 to 17 inches, while the ball is somewhere from 4 to 6 feet away. Lineless lenses or progressive lenses can mitigate much of the focus problem—the blurriness of the ball—but not all of it. As you shift focus, your sight lines move through areas of the lenses with different focal-length attributes. The problem arises from not only having to make quick focus changes for different distances, but also making these changes while peering through different focal lengths.

3. **Finish the visualization with playback.** Replay everything to help reinforce what can be learned from the stroke for future use. Visualize the complete sequence of events that led to either

a make or a miss, and all the images that you centered upon during the process. Try to make this a part of your routine for all shots, good as well as bad. For poorer efforts, you will be analyzing what went wrong and when the mistake was made, as well as what the consequences were, and how the error can be avoided in the future.

It's important to note that the visualization exercises used by professionals are more complex and detailed than a beginner's. But as your game develops and becomes more sophisticated, your visualizations should become more sophisticated too. The more you do something right, the more it's ingrained in your motor memory. Eventually, practice is carried out more unconsciously, and your attention can be turned to adding complexity to your visualization.

The picturing of the golf swing uses many of the same neural pathways as actual vision, and those pathways are strengthened when exercised by visualization. Visualization makes the connections between nerves and the muscles they control grow stronger, from the motor cortex of the brain outward to the peripheral nervous system, just as visual imagery strengthens the neural pathways in the opposite direction (that is, those emanating from the peripheral nervous system and leading back to the brain). As a result, the muscles actually have the potential to generate greater velocity and power without engaging in any strength training. Nothing changes in the muscles themselves; instead the fortified circuits in the nervous system make it possible to get more from the muscles.

Visualizing entire shots and continually adding details as your game develops are very appropriate for the intermediate to advanced player, but not necessarily the rest of us trying to correct bad habits, which have a tendency to become burned in our memories. Even pros have difficulty correcting bad habits, so don't get discouraged. The best way to get out of these ruts is to visualize related activities such as the snapping of a whip or towel, as described in Chapter 2.

Activate Kinesthetic Memory

A parallel or complementary performance aspect to visualization is kinesthetic memory—the memory of appropriate musculoskeletal movement. This process, by which a golfer's muscles and joints move in a familiar manner, conditioned by training and past experience, is the result of tactile, visual, auditory, and inner ear sensory information compiled through years of practice.

When picking up a golf club, our hands instantly begin relaying information to the brain, activating kinesthetic memory. The motor cortex then responds by exerting a force on the club, registered as the tightness of the grip. Thus, the hands are used to explore, activating kinesthetic memory, and then they manipulate the club, executing directions from kinesthetic memory.

Before beginning the backswing, professional golfers often engage in a full-body shaking motion; it looks like he has a case of the jitters. Among the golfing crowd, it's referred to as a waggle, and this movement, which is often mistakenly interpreted as nervousness, serves the same purpose as a basketball player bending his knees, dribbling, and rotating the ball across the fingertips before a free throw. The basketball player uses a preshot routine and the golfer uses a waggle because both athletes begin movement from a static position and share the goal of trying to precisely coordinate the movements of hundreds of muscles in order to sink a ball in a hole. Basketball coaches often chastise players for coming in off the bench cold and immediately launching a shot. They were in a static state on the bench and did not have a chance to get a feel for the ball and their shot by going through the more simple body movements involved in the game before taking on the complex task of shooting. Similarly, to establish rhythm and timing and get the muscles moving in the right pattern and sequence, golfers need the waggle.

With enough practice, your kinesthetic memory can be trained in quite remarkable ways. Ted St. Martin, who never played college or professional basketball, set a world record of 5,221 consecutive free throws in a row, and he has such confidence in his kinesthetic memory that he is capable of sinking over 50 free throws in a row while blindfolded! Meanwhile, the muscle memory necessary to accomplish this very repetitive task is activated completely by the feel of the ball rolling across his fingertips.

Although the same principle holds for hitting the golf ball, the golfer must execute in a far harsher environment. The variety of golf strokes needed through a round and the day-to-day changes in course and weather conditions often make it necessary to search the recesses of our memory for similar shots made under comparable conditions. Memories of quite similar but not identical shots create a mental tug-of-war between doubt and confidence in the brain. In the end, the confident golfer feels assured that he has the right motor memory picture, while the doubter can't decide which among three or four images is best in this situation. So do not overwaggle: It only takes a few seconds to activate motor memory, so waggle until the right picture pops in your head, lock in on that picture, and swing away. Anything more than two or three seconds of waggling is unnecessary and distracting.

To activate your motor memory consistently, go through the same ritual before every swing: a fluid practice swing, a couple of pats of the heel of the clubhead against the turf, and a few seconds of waggle before firing.

Fully developing your sense of touch is a vital aspect of improving kinesthetic memory. The majority of touch receptors are located in the fingertips, which are incredibly sensitive. Since each fingertip has over 2,000 receptors just for touch, if the club is pressed up against the palm instead of the fingertips, you are losing out on valuable sensory feedback. Of course, the palm has touch receptors too, but the density drops off significantly at the base of the fingers and palms. Thus, it makes sense to move your fingertips around and squeeze and relax your grip a couple times as part of your preswing ritual.

Practicing the same stroke over and over is the only way to establish kinesthetic memory. After the muscles, joints, and tendons have adapted to executing the 5-iron swing flawlessly, the brain's "wiring" has been established, but this practice does little to help improve your skill with the pitching wedge. That is why many instructors will tell a hacker who wants to achieve the lowest score possible in the shortest amount of time to carry only four clubs (a 3-wood, 5-iron, wedge, and putter). There will be a smaller range of options and less potential for remarkable shot-making, but this is more than offset by the stronger kinesthetic memories developed with the clubs that are carried, which should result in a much better overall result.

Besides all the kinesthetic memory involved in the thousands of muscle actions during the golf swing, you want to develop a feel for the contact. The kinesthetic memory of contact includes the feel for the right vibration when hitting the sweet spot and the right interplay of the clubface with the ball for extending impact and imparting the desired spin.

Learning to master all the clubs in your bag requires that you highlight your past achievements. Dwell more on the kinesthetic memory of what you did right and avoid an extensive evaluation of what you did wrong. You need to accentuate the positive, retaining the memory of great shots and forgetting the bad ones. This elective memory approach will help you gain confidence in your clubs as well as better and more varied skills. You cannot expect to instantly hit draws and fades and stick greens like lawn darts, but you can become much more proficient by improving your kinesthetic memory with practice.

• Ray Floyd on Memory Activation

RAYMOND FLOYD eloquently and anecdotally describes his kinesthetic memory activation process this way: "A scorer is obsessed with the target on every shot. When I'm playing my best, I immerse myself in my target. As I go through my preshot routine, I decide where and on what path I want to hit the ball. In my case, once I get over the ball, I have a kind of rocking action with my feet that helps me filter that feel for the target through my body to my hands. When I feel at one with the target, that's when I pull the trigger" (*Golf Digest*, February 1999, p.119).

Confidence

There is really not that much difference in skill level between the guys at the top. Top professional golfers are pretty much at the same level physically, so the separation in their ranks is largely decided by who has the confidence and mind control to keep all the thousands of intricate and subtle actions of the joints and muscles in harmony. That's why it's often said that confidence, the full trust in their skills and clubs, is what elevates the best above all the rest.

It is important, however, to dispel the widespread misconception that confidence is tied to great conscious thought about mechanics. In reality, confidence entails less conscious thinking about mechanics, not more. A confident free-throw shooter does not lecture himself on the step-by-step motion of the shoulders, arms, elbows, wrists, and fingers, and neither does a confident golfer; the action happens too fast for step-by-step instruction. Instead, he zeroes in on the objective and allows instinct to take over, performing without interference from the conscious mind, and once he finds this comfort zone, he wants to do everything he can to keep repeating it.

Confidence is not something you're born with; it's something that you have to work at—continually improving visualization and kinesthetic memory through practice. You must be confident enough to believe in your analysis of visual and touch sensory information—visualization of the stroke and kinesthetic memory—which is strongest for the shots that you practice the most. You want to be secure in the belief that the 5-iron you pulled from your bag will carry the intended 180 yards, not 170 or 190, just like it did in practice. Thus, you are much better off practicing in such a way to develop greater confidence in your existing abilities than concentrating on expanding your capabilities. However, it is equally important that you're honest about what you're capable of doing; you don't want to overestimate your abilities.

Even for the best professionals, there are different levels of confidence with different clubs, and because your skill varies by club, your strategy should as well. If you're only mildly confident that you can reach a green with a 3-wood, but certain that you can lay up perfectly with a 5-iron, definitely opt for the 5-iron. Always select your club confident in the probable outcome; the best-case scenario with the 5-iron may not be as good as that with the 3-wood, but you're far more likely to realize that best possible outcome. You never want the opposite: an extremely aggressive strategy you don't necessarily have the ability or confidence to execute, with an unsatisfactory probable result. The knowledge that the odds are not good that the shot will be properly hit adds another layer of indecisiveness to execution, which in turn further lowers the odds of a good outcome.

A consistent approach to every shot (one that activates kinesthetic memory), practice that develops confidence in your clubs (strengthening and reinforcing kinesthetic memory), and a realistic evaluation of exactly what your capabilities are with each club are the three most important ways to eliminate doubt and second-guessing. If you're swinging with doubt in your mind, there is no way you can expect flawless motor control; your mind believes it's beyond the realm of possibility.

Besides the psychological reasons why second-guessing is detrimental, a last-second change of plans exceeds the abilities of the nervous system. You need to execute decisively with your preswing strategy because it takes a minimum of one-tenth of a second between visual processing and adjustments in motor action, which is about the time from the start of the downswing until contact. It's the same reason you should not keep a checklist of tips to follow when swinging: Incorporating tips and things learned during lessons should be left entirely to the practice tee, and even then it's best to work on only a few things at a time. Otherwise your brain becomes too cluttered to perform effectively, and no lessons can be internalized or applied.

• Maintaining a High Level of Concentration

A LUXURY of golf is that you do not have to focus for four hours straight; you only have to concentrate intently when you start to prepare for a shot. Further, there is no time constraint; you can take as much preswing preparation as you like. Unlike tennis, where competitors have to be highly focused throughout the entire match (except for the short break for side changes) as they continually read, react, and execute. That's why so many professional tennis players retire in their mid- to late-twenties; they are still in their physical prime, yet they claim mental fatigue, citing the increasing difficulty of maintaining that high level of concentration. In golf, the luxury of time allows for the visualization and kinesthetic memory feedback processes of the mental game to take place at whatever leisurely pace you desire.

Slumps and Choking

Occasionally, a PGA golfer gets into such a groove that his name becomes a fixture near the top of the leader board tournament after tournament. Then, suddenly, the near flawless play mysteriously ends, and the golfer can go several years without a single victory or even a top ten finish.

There's no universally applicable explanation for slumps of this nature. If there were, they would never occur; players would quickly go to the "formula" and make the necessary adjustments to start playing well again. Even with the tremendous prize money in golf today, which makes it possible to hire the best coaches, their advice often does not help. Players get mired in slumps, and regardless of the advice from swing doctors, they can't seem to rescue a faulty game.

Underlying any slump is a loss of confidence. Sometimes it begins due to a slight flaw in a golfer's swing mechanics, but in trying to correct the glitch and overcome that distraction, it's very difficult to reestablish the confidence that the swing is indeed "correct" again. Players only break out of slumps when they again feel confident in their abilities.

Whereas a slump is an ingrained, long-term confidence problem, choking is a short-term breakdown in confidence—a momentary doubting of motor memory—and doubt leads to performing more consciously in a step-by-step manner what ought to be done subconsciously. In other words, it results in thinking more about mechanics instead of less, and a distrust of motor memory. Furthermore, muscle tension factors into the equation. Sometimes with increased doubt comes significant muscle tension, which throws off execution; other times the pressure of the situation creates the greater muscle tension, affecting the feel for the shot. Your muscles do not feel quite right, so neither can your swing feel right.

The best way to avoid choking is to continue to breathe normally and avoid altering the visualization and motor-memory processes. Before going into the preswing routine, you want to feel the familiar kinesthetic feedback from the muscles in your arms—if they feel tighter than usual, take a few deep breaths and shake your arms a couple of times to relieve excessive tension.

The Zone: Real? If So, How Do I Find It?

You see it often: Under the enormous pressure of a final round of a major championship, an incredibly hot golfer drops in several chips and putts from great distances. Tom Watson won many major championships in the late 1970s and early '80s by turning in very memorable Sunday afternoon shot-making heroics, and his uncanny chipping and putting exploits were attributed to his being in "The Zone," an uncanny can't-miss hot streak.

Chipping and putting are skills learned by imitation, trial-and-error, and constant practice. Shot-making is a marvel of human kinetics. For putting, it requires the golfer to consider the correct pathway and correct launching speed for a variety of distances, breaks, and conditions. In tournament situations, under pressure, a hot player will drop almost all putts at distances of less than 10 feet, and the majority between 10 to 15 feet.

It seems unbelievable that the human body can be trained to reproduce the required movements so accurately, but good kinesthetic memory allows golfers to recall and duplicate the shot. Although all professionals develop this memory through practice, there remains a wide disparity in shot-making prowess. Some of this gap can be partially attributed to differing skill levels, but a significant proportion can be attributed to diminished confidence—an inability to overcome self-doubt, manifested in the questioning of visualization and strategy. Thus, The Zone may be explained as nothing more than a relaxed state of concentration and heightened confidence. While in The Zone, golfers comment about experiencing a dual sense of state: a heightened level of mental concentration, yet a very relaxed muscle state. This state makes it possible to excel at all four components of the mental game: an extreme confidence in the visual analysis, the ability to visualize a clear mental picture of the swing, total activation of kinesthetic memory of the muscles and joints, and the muscle relaxation that ensures that swing execution matches the brain's commands.

It seems counterintuitive that someone in The Zone is thinking less rather than more, but that is indeed what happens, and it can be explained by physiology. Scientists divide motor movements into three groups—reflex, voluntary, and automatic—and when in The Zone, voluntary motor movements become more like automatic motor movements. In other words, once a movement has begun, and then repeated many times, the action can continue almost automatically, with no conscious thought, until the movement is stopped. A marathon runner, for example, does not have to think about moving his arms and legs for three hours straight, they move automatically. Of course, the golf stroke is not as consistently repetitive as the motion of a runner's arms and legs, but professionals who describe experiences in The Zone usually mention the feeling that complex athletic moves were performed automatically.

In *The Game I Love,* Sam Snead offers one of the best descriptions of The Zone: "When you're in the zone you feel more relaxed. Everything feels smooth. Your senses become sharper. You see all things more clearly. You can see the line of every putt. Your visualization is clear . . . You're not trying to hit the shot, you just do it. And there are no mechanics—heavens, no, there's nothing to think about. There's work to do, and you just do it and go to the next shot. You never feel as if you can't do what your mind is telling you to do."

The hope is that someday the scientific community will provide instructions on how to find The Zone. It has already been tried by attaching electrodes to the scalp to monitor the very faint electromagnetic emissions (about one-millionth of a volt) that the brain sends out in order to identify an emission pattern for someone in The Zone. One such experiment involved monitoring the brains of free-throw shooters to discern distinguishable patterns in the flow and rhythm of alpha-wave patterns (the brain's electric wave activity). However, the problem with that test and other studies hoping to pinpoint a certain pattern to the coveted Zone is that sometimes electroencephalography patterns (or EEG patterns, referring to the machine used to record alpha waves) respond predictably to certain external stimuli, while other times the brain-wave patterns become partly obscured by "noise" (interference) from other brain activities.

If the alpha-wave patterns associated with The Zone can be pinpointed, someday coaches may be able to train golfers to zero in on it. Perhaps someday brain-wave monitoring equipment may be marketed to instructors as a surefire way to help golfers reach the "Holy Grail" of the mental game, or just break out of a putting slump. Until that day comes, however, probably the best remedy for a slump is to try to create the focused mind and relaxed muscle state that's associated with The Zone.

Putting It All Together: Five Things to Avoid

Now that you are aware of the key aspects of the mental game and how to improve your mental approach, it is worth mentioning a few ways golfers overthink—too much of a good thing—that can prove detrimental to your game. In short, if you try too hard to emphasize the mental game, you may do more harm than good. The following are five keys to keep in mind to ensure your cerebral approach to the game stays on track:

Don't Overpractice

You hear it all the time, "practice, practice, and more practice is the only way to improve." This advice can be true, but it can be terribly misleading as well. For an aspiring professional golfer, the quality of practice is more important than the quantity.

Neuroscience research has shown that it takes a high level of concentration while practicing to develop and cement kinesthetic memory for the perfect golf swing. The motor system is composed of billions of neurons (tiny nerve cells) that are organized in what are referred to as "maps." These neural roads and highways, which respond to sensory inputs, represent our skills and knowledge of the world and are where the kinesthetic memory of the swing is stored. When a golf skill develops or changes, the maps also change: Neuron populations undergo a reorganization to execute a specific skill, but if these neuron maps are poorly developed (because there hasn't been enough practice reinforcing the skill), they can be abandoned.

Practice that is approached casually makes no progress in the rewiring effort. Since there are tug-of-wars going on all the time in the wiring of our nervous systems, our brains have a built-in protection feature to not change unless an investment has been made to practice intensely. The brain is a limited resource; it has to be selective in what it decides to keep and what to disregard, and it takes serious practice, a lot of repetition, attention to detail, and a deep level of concentration for the brain to exhibit the plasticity to rewire the nervous system. By practicing golf skills with special attention and intensive repetition, the chemical messages cement a lasting increase in the connection strength between the neurons, which positively impacts kinetic memory. For those with exceptional skill whose goal it is to improve further, the brain maps that already exist can become tighter through practice, which leads to a better synchronization of the neuron actions. And the tighter your brain maps, the higher your confidence will be.

As wonderful as this receptiveness to learning can be, there is a drawback to the phenomenon as well: When one map grows, another might shrink. Since maps compete with one another for neurons and space, the development or growth in, for example, the map that governs the execution of a flawless short chip-and-run shot might come at the expense of that pertaining to use of the short wedge. Therefore, even when focusing on improvement of a very particular action, you always want to be practicing similar skills as well so that any improvement in one skill set does not come at the expense of another. A good example is the 2000 Summer Olympics female gymnastics all-around event: Mysteriously several of the competitors fell while trying to execute one or both of their vaults. Analysts claimed they were cracking from the pressure of competition. Only later was the real reason revealed: Someone incorrectly set the vault horse a couple inches lower than it should have been, and those few inches were all it took to completely throw off the gymnasts' mechanics and make some very good athletes look very bad. If vaulting entailed two vaults from two slightly different heights (just as golf entails ever so subtle differences in swing dynamics for very similar shots), the gymnast working on improving her weaker vaulting technique from the slightly higher elevation must not forget about mixing in a couple of "reminder" vaults from the lower height to ensure she can distinguish the two slightly different dynamics and complete the routine. Likewise, a golfer practicing 50 short chips from the fringe of the green on an uphill slope to the cup should also take a few practice strokes from deeper and shorter grass, from slightly different distances as well as from downhill and uneven slopes to the cup. Your motor memory best develops the confidence to distinguish slightly different strokes by including a few of these "reminder" strokes into any practice session.

Equally important to how you practice is the quality of practice. An aspiring professional golfer will get far more value from an intensive highly focused half hour of work than a casual four hours. Just because a golfer practices four hours a day does not mean he is getting more value from his practice. Although there are golfers who can remain alert for hours on end, most aspiring pros would be better served by breaking their practice time into two or three sessions a day so that it is much easier to maintain the desired focus for the entire session. If you approach practice casually or spend two hours concentrating intently on your game, followed by one hour of casual training, you might be doing more harm than good. Practice is a lot like road building: You better do it right—with a high level of attention—or the finished neural roads and highways in your brain will be faulty, and some of the faults will be deeply, irreparably flawed.

Don't Overrely On Your Eyes

Trust your eyes, but don't overrely on them. Your senses improve by working them, and the more you depend on your eyes, the less your other senses are called upon. For example, the blind usually have far sharper senses of touch and hearing than the rest of us. In the absence of sight, the other senses compensate for the handicap, and the blind become more acutely aware of the sensory input of touch and hearing.

An excellent way to improve motor memory is to practice putting with your eyes closed. From five feet away, start putting from the exact same spot until you drop two or three in a row. Now try the same putt, but close your eyes before striking the ball and do not open them again until you think the ball should be within inches of the cup. This drill will sharpen your feel, your ears will be more alert for the right "ping" on contact, your arms will develop a feel for the speed of the pendulum motion, and the touch receptors in your fingertips will be more confident for the correct feel of contact, especially the sweet spot and time of contact.

Don't Overfocus

Staring at the ball too long before the stroke tends to lead to mishits because the ability of the eyes to center on a small target like the golf ball diminishes over time. Do not fixate on the ball too early, just grip and rip. Use a soft focus on the ball until you're set and ready to swing, then switch to a hard focus when you begin the backswing.

Don't Move Your Head

Although almost all the major joints and muscles feel a call to action, you want to keep your head as stationary as possible throughout the swing. The physiological reason here is simple: It is hard to execute properly if there is an earthquake at your command and control center. Movement disrupts your vestibular system, which is located within your inner ear and which controls the body's balancing mechanism, best known to most of us when it becomes disoriented and causes seasickness. This doesn't mean that moving your head with the rest of your body makes good swing mechanics impossible, simply that it will be much more difficult, because head movement tends to disorient you in relation to the ball. A more dramatic example of the effect of head movement on performance occurs when a football receiver has to sprint all out and look overhead to catch a long pass. Because his head is bobbing up and down as he stretches his neck muscles back to look for the ball, the catch is extremely difficult. Though we regularly see the catch made and know that it can be done, nevertheless, it's an extraordinary athletic feat.

Another reason to keep the head steady is to reduce the number of visual skills needed. When your head is stationary and you're trying to cleanly hit the stationary ball, the primary visual skill employed is static acuity. If either the head or the target is moving, it requires far more visual skills, such as dynamic acuity, eye tracking, and accommodation (the ability to change focus quickly as the distance from the target changes). And as any baseball centerfielder trying to run down a long fly ball will tell you, perfecting these visual skills takes a lot of practice time.

Don't Overthink

Contrary to popular wisdom, the best pros don't think more than you or I, they think less. From years of practice, the voluntary rhythmic movements become automatic like the reflexive knee bend when a doctor taps on the patella tendon below the kneecap.

PGA professionals don't have to think about their mechanics; they develop a natural rhythm and timing that's so automatic that they don't clutter their minds with thoughts about their execution. Once the backswing has begun, the rest of the swing is carried out almost effortlessly, like a reflex, until it's finished. The golf swing for a pro is as natural for them as walking is for us; they don't put any conscious, step-by-step thinking into swinging a golf club, just as you don't actively dwell on the mechanics of walking. Unfortunately, most of us will never develop that kind of proficiency since we can't devote the practice time necessary to improve so dramatically. Because there is little chance your game will ever improve if you maintain a mental checklist of 100 things you need to account for before every swing, the better option is to seek incremental improvements by limiting the number of things you are working on at any one time. Train your swing on the practice tee and then trust it on the course. If you dwell on swing mechanics as you play, you will not score as well as possible. Overthinking leads to second-guessing, which invariably erodes confidence.

Great Golfers: Nature or Nurture?

Golf is one sport where gamblers wouldn't have much luck picking winners by looking at the various athletes. There is not an identifiable physical profile that is better than any other, and there probably won't ever be regardless of what genetic research uncovers. Of all the major sports, golf is perhaps the closest to offering an equal opportunity of success to all comers.

There's no doubt that human structural features have the potential to dramatically constrain or assist athletic performance in sport, but golf is the exception to the rule. Genetically, great golfers do not possess specialized physical traits like sprinters, high jumpers, and long jumpers do. Whereas sprinters and jumpers have a higher center of mass, less body fat, narrower hips, thicker quadriceps, longer legs, higher calves, wider shoulders, and a per-feet gait (they can't possess physical defects such as bowlegs or pigeon-toes that would hamper energy efficiency or speed), there is immense variability of physical attributes found among PGA golfers. Track and field athletes are also extremely specialized genetically in regards to muscle makeup (the proportion of slow-twitch to fast-twitch muscle fiber), while that trait is a virtual nonfactor for golfers. Scientific studies of sprinters and jumpers have shown an incredibly high proportion of explosive fast-twitch muscle fiber (up to 95 percent), and while having a greater proportion of explosive fast-twitch muscle fiber is certainly beneficial for the power game in golf, rhythm and timing are far more important. Moreover, although science has found anaerobic power to be anywhere from 40 to 90 percent hereditary, there is still a trainability aspect to muscle makeup that varies from person to person and can be strongly influenced by early childhood training. Trainability remains a big question mark; it is still unknown how much it affects physiological factors. Some studies have shown an initial 5 percent difference in anaerobic and aerobic traits between two individuals widens to 15 percent to 20 percent once both subjects begin following identical training regiments.

Although success in golf is not predictable from physical attributes, it is highly dependent on motor control. Unfortunately, knowledge of the hereditary aspects of motor control is still in its infancy, and only a small fraction of the processes that take place in the nervous system governing motor control is known by science. Since studies have found significant variability in how physiological performance improves when subjected to training, it only goes to reason that heredity plays a major part in how quickly and effectively the nervous systems of golfers adapt to motor control training. Unfortunately, the trainability question may never be answered by scientific studies because it is impossible to link the developmental progress with the training. Two eight-year-old beginners are not starting out equal, because no two children have identical developmental environments. For example, who is to say whether the transferability to golf of the eye-hand coordination of the eight-year-old who played baseball is better or worse than one who played only soccer? Even if there was some magical way to equalize the starting skills and developmental environment, you can expect many different scenarios for a beginner picking up a golf club for the first time: the natural who goes on to be a scratch golfer; the natural who quickly reaches a performance plateau, never breaking 80; the quick study who advances at an accelerated pace and keeps improving; the quick study who plateaus early after an initial accelerated pace; the challenged who initially develops skills at a slower pace but eventually becomes a scratch golfer; and the challenged who exhibits very limited proclivity for the game despite endless hours of practice and lessons.

Because of these extremely unpredictable results, it may never be known why there is such a wide range of responses by the nervous system to training. But there is one sure thing: All professional golfers have very trainable nervous systems. Great golfers are made, not born; they are fashioned by the nervous system perfecting and automatically performing the very specialized and subtle movements of the golf swing. Obviously it helps to start young, and except for a few notable exceptions such as Se Ri Pak, the LPGA 1998 Rookie of the Year who did not start playing golf until the age of 14, almost all professionals started playing golf at a very early age when their nervous systems still had their greatest plasticity—the exceptional ability of the young nervous system to change the efficiency and strength of the connections between neurons. And the greater the plasticity of your nervous system, the greater the opportunity to perfect and ingrain the motor control skills necessary to excel.

In an increasingly competitive world, the trend is for athletes in many sports to begin training at younger ages while the nervous system has its greatest plasticity. For golf this is particularly true since it is a sport that requires fine-tuned motor control skills. The success of Tiger Woods, who began swinging a golf club by the age of two, has many parents signing up their preschoolers for golf lessons. With the best instruction, can these kids turn out to be scratch golfers? You bet. Can they eventually be good enough to join the PGA Tour? You never know.

Should You Be Golfing Left-Handed?

Think about the advice given for reaching swing perfection: Let the left side direct the backswing; let the left side initiate the pulling motion in the downswing; let the left side generate most of the power for the swing. With all this work being done by the left side, the right side of the body is more or less just a fellow traveler with the club, adding very little power to the equation and only marginally controlling the action. Perhaps it's more than a coincidence that many of the greatest golfers of all time, such as Jack Nicklaus, Ben Crenshaw, and Tom Kite, are left-handers who took up the game of golf from the right side. They write left-handed, throw left-handed, and are left-eye dominant, yet all swing a golf club from the right side. And Phil Mickelson, the number-two golfer in the world in 2001, is the opposite: a right-hander who plays left-handed.

Considering the success of these professionals, you may be wondering if you should try cross-swinging? Logically, the dominant side of the body—the side with greater strength and motor control—should be the side that is called upon to do most of the work. However, as logical as this reasoning seems, there are many reasons why most of us don't cross-swing, and just as many reasons why we shouldn't change now.

Before delving into the cross-swinging question, it is useful to consider what dominant handedness is all about. Brain organization and activity determine which hand predominates, with each hemisphere of the brain processing motor and sensory activity on the opposite side of the body. And although the two hemispheres look nearly identical, their structures and, more important, their functions, are not. In particular, areas responsible for motor control are generally much more highly developed in one cerebral hemisphere than the other, and this is called the "dominant" hemisphere. For 90 percent of the population the left hemisphere is dominant, which explains why the overwhelming majority of people throw with the right hand and kick with the right foot. In the other 10 percent—lefties, more affectionately referred to as southpaws—the right hemisphere is dominant.

What "makes" a lefty is still unknown, but contrary to popular belief, scientists believe that hand dominance is not strictly inherited. Although they widely agree that genetic factors play a very important part in that determination, there are several factors during the course of development, both pre- and postnatal, that affect the direction and magnitude of the genetic differences. It seems that all humans carry the genetic code to be right-handed, yet something happens during development that makes an individual left-handed. Up until infants reach about two years old, scientists report that they frequently use both hands interchangeably. But as they begin to perfect fine motor skills, a dominant hand emerges. Establishing a dominant foot takes a while longer and is determined usually around four or five years of age or even later. Perhaps this occurs less because of physiology and more because we do not need to develop fine motor control skills with our feet-walking and running and even the slightly more complex task of kicking only require gross motor skills.

What perhaps makes the golf swing so difficult to master is the fact that the body must simultaneously perform a number of fine and gross motor skills. During the maturation process, fine and gross motor skill development proceed independently, as the brain learns to deliver precise motor control instructions to the muscles. In order for that to start to happen, billions of neurons in the body must develop a myelin sheath, which is an insulating material that serves to speed nerve conduction. Concurrent with the neurons forming a myelin sheath, the synapses (the junctions between neurons that act as the brain-to-body communication channels) begin to create efficient and well-organized networks out of the random chaos of billions of nerve cells.

It's akin to designing and building a railroad from scratch—the more direct the lines and efficient the switching stations, the quicker the transportation. And there is considerable evidence that the earlier we start to "build" the nervous system—that is, learn fine motor skills—the better our skills will be since the nervous system of an infant or toddler has greater plasticity. In particular, repetitive and continual practice results in structural changes in synapses and greater concentrations of neurotransmitters, which in turn makes nervous system activity more effective and efficient in carrying out precise motor activities.

The vast majority of players today learned golf as a teenager or older and most spent much of their youth playing other swinging sports like baseball and hockey. When the time came to pick up their first golf club, it was only natural to assume the same sideswing used in baseball. The visual cortex of the brain draws upon kinesthetic memory from those past similar experiences to apply to the present situation in trying to best execute what is unfamiliar. After comparing the first golf swing experience against those from baseball, information is relayed by brain cells (neurons) to the motor cortex, which controls the muscle movements necessary to swing a club. Neurons receive, analyze, and transmit information through the body's synapses to the muscles, and practice helps synapses grow stronger, allowing swinging skills to dramatically improve. However, in many ways this also works against the baseball player taking up golf because his synapses grew stronger from the practice of swinging a baseball bat so that the right-handed swing became very natural, comfortable, and fluid. Although it may seem that the fluidity of the baseball swing would be a boon, that's not the case, primarily because for a right-handed hitter, the right arm provides power and control while swinging a bat, but with a golf club it's the left arm that provides most of the strength and stability. Thus, the "neglected" neurology controlling the left side did not reach its full potential because of a lack of practice.

This is not to say that it's impossible to become a good cross-swinger, it just makes it more difficult to develop the confidence to trust kinesthetic memory to perform the fine motor skills needed. Excelling from the right side is perhaps an even greater challenge: It requires a conscious effort to develop a golf swing from scratch since many baseball swing characteristics are detrimental when applied to golf.

For most of us, it isn't easy to "retool" the synapses to play the game from the left side. If you have been playing golf right-handed, it makes little sense to switch to left-handed clubs now. In the link between age, motor development, and performance, there is such a thing as "too late." Experts generally agree that the brain of a child up to eight years of age still possesses significant plasticity to train the visual and motor system. After the age of eight, it becomes progressively more and more difficult.

Tiger Woods, a right-hander swinging from the right side, began swinging a golf club as a toddler. Although his father chose right-handed clubs for his right-handed son, there is little reason to believe that he would have been better if he began golfing left-handed. His early start meant the plasticity was there to be proficient from either side, and there wasn't any ingrained motor control habits from other sports to overcome. The fine motor control needed in golf and the very competitive nature of the PGA Tour, largely explains why the majority of professionals began playing the game at a very young age, and why none of the greatest athletes in other sports have been able to take up golf and reach the PGA Tour. Of course, there are a few exceptions, like Babe Zaharias, who excelled in numerous sports besides golf-where she managed to win 17 amateur titles in a row from 1946 to 1947, followed by 31 professional tournament victories.

Unfortunately, what sounds good in theory has not been tried much in practice. Few beginners consider the cross-swinging option, and when they do, it is usually lefties considering righty clubs. But this has more to do with the market for golf clubs than anything else. There are simply far more righty clubs out there, and most golfers start by borrowing the right-handed clubs of a friend or relative.

• More Left-handed Athletes?

SINCE lefties are "wired" differently, there is widespread speculation that they have an inherent advantage to excel in sports. Statistics seem to show that the number of left-handers in sports closely reflects the 9 to 10 percent found in the general population, but in some sports—especially those that require quick reflexes, such as boxing, fencing, and tennis—there regularly are much higher percentages of lefties found among the top ranks.

• Defining the Dominant Hand

SOMETIMES it is difficult to define which hand is dominant. For example, Steve Flesch, who had 13 top-10 finishes in 2000, throws, writes, and began golfing right-handed, but switched to left-handed clubs at the age of 10.

4

Getting the Ball from Here to There

THE essence of golf is getting the ball from here to there with a varying repertoire of power and precision—power to a general area, precision to a specific target. Of the two, though, precision is far more crucial; regularly launching 300-yard drives is a noteworthy benefit of limited practical importance. Power comes in handy off the tee and for second shots on long par-5s, but these usually make up about 20 percent of the strokes that go into any round of golf. Further, an accomplished power game can actually do more harm than help. Since a distant drive gives a big hitter a head start on most holes, it's easy to develop a false sense of security and be complacent about improving the precision game. This addiction to power at the expense of precision, though prevalent, is indeed foolhardy: Whether you leave the ball 20 yards or 50 yards from the green, it still requires excellent touch to get the ball where you want it to go. As long as the rules of golf prize dropping a ball in a cup in the fewest strokes possible, the probability for success will favor those who master the short game—the players who are able to get the ball to a very specific target.

The visible aspects of precision shot-making are indeed amazing and thrilling to observe, yet true connoisseurs of the game are more impressed by the underlying subtle and invisible stuff going on as well. Specifically, they marvel at how spin is imparted to the ball, and how its trajectory is altered as it sails through the invisible medium of air. To develop a deeper appreciation for the invisible and subtle, this chapter looks at shot-making from two perspectives, examining both human-controlled collision factors and environmental ones. The environmental factors involve how the ball in motion—while in flight, bouncing and rolling—is affected by the environment; the human aspects pertain to the collision dynamics—how one controls and manipulates the interplay of the clubface and ball so it travels in the air and on the ground as intended.

Flight, Bounce, and Roll

Since golf is a target game, the ball must travel from tee to hole through some combination of flying, bouncing, and rolling; how much or little of each component is up to your discretion. The greater your skill, the greater your options; in other words, good golfers do not make shot-making decisions by physics alone, as ability always trumps physics. For example, if you lack the confidence or skill to hit a bump-and-run chip shot, you shouldn't even try, even when it is a sounder shot-making strategy from a physics perspective. A well-executed lofted shot that you are confident in will usually give you better results than a poorly executed bump-and-run shot even when the latter is strategically superior.

Ball Flight

On its way from tee to hole the ball travels the majority of the way in the air. The forces that determine how a golf ball will travel while airborne are the initial velocity, launch angle, gravity, and air resistance, with the latter two being highly dependent on environmental conditions.

ENVIRONMENTAL FACTORS

Gravity is the most profound force acting on the flight of the ball, yet since it is a constant one—9.8 meters (about 32 feet) per second, per second (m/s^2)—it is given little thought. We develop a familiarity with the constant of gravity at a very early age. As a child first learning to catch, we not only were learning the physical skills involved in entrapping a ball, but also how to judge the way gravity affects its flight, which is actually the tougher of the two skills to learn. The most important part of catching is to get the hands in a position where they can grab the ball; we have to judge where it's going so that our brain can position our hands to make the catch. In golf we do not have to catch the ball, but we do need to visualize how gravity will affect where the ball will go.

Wherever and whenever you play, gravitational acceleration exerts a fairly consistent force throughout the ball's flight path, on the way up as well as on the way down. "Fairly constant" is an important qualifier here since gravity does vary significantly at different latitudes and altitudes. Latitude variations are due to the shape of the earth: since the earth bulges at the equator and flattens a bit at the poles (a difference of 18 miles or 29 kilometers between its equatorial radius and polar radius), you experience 0.18 percent less gravitational force at the equator than at a pole. There is also a latitude-related variation due to a difference in the centrifugal effect of the earth's rotation—the outward-flinging effect of anything moving in a circle. Because the rotation is greatest at the equator and nonexistent at the poles, gravity is reduced the most at the equator and not at all at the poles—a difference of 0.35 percent. In total, the two effects combined—the centrifugal effect with the effect of the earth's nonspherical shape—reduce gravity by 0.53 percent between the equator and a pole. These variations might not seem like much, but the difference in distance that a golf ball will travel can be substantial, especially when the course is located at a high altitude, because as you move farther from the earth's dense core, gravity becomes a weaker force, decreasing at roughly 0.0066 feet per second squared (0.002 meters per second squared) for every kilometer of elevation.

Aside from gravity-related changes due to altitude, there are also discrepancies in air resistance. Atmospheric (air) resistance is proportional to air density, which decreases at higher altitudes and higher temperatures. For example, at an elevation of one mile above sea level, air density drops by 17 percent from its value at sea level. The combination of less gravitational acceleration and lower air resistance is what makes mile-high Coors Field in Denver a home-run paradise for baseball hitters, as long fly balls carry over 5 percent farther than at ballparks near sea level (a 340-foot long fly ball in Yankee Stadium is a 357-foot homer at Coors Field). The same dynamics hold for golf. Physicists Frank Werner and Richard Greig, the authors of *How Golf Clubs Really Work,* calculated that the drop-off in gravitational acceleration from sea level to 6,000 feet would add 11 more yards to a 209-yard drive. If you consider gravity variations due to latitude, as well as other aerodynamic benefits when the ball is struck at higher velocities, the same initial launch angle and velocity that results in a 300-yard drive at Saint Andrews on the coast of Scotland (56 degrees latitude, and a few feet above sea level) could easily result in a 330-yard drive in Denver (40 degrees latitude, 5,000-foot altitude), or a 350-yard drive near Quito, Ecuador, located on the equator, with 8,000-foot altitude highlands.

If a drive sails 330 yards instead of 300 yards, it's indeed helpful in lowering your score for any given round of golf, but the critical concern is the effect of gravitational variations on long-range precision shots, such as medium-to-long approach shots. For example, the 200-yard 3-iron shot that drops a few feet from the hole in Scotland will be a 220-yard ball that sails well past the green in Quito. An extra 50 yards of drive distance is of little benefit if you overshoot the green by 20 yards with the second shot.

Altitude and latitude variations in gravity, which have the potential to wreak havoc on the precision game, seldom have any effect on a wise golfer. Well-prepared pros don't arrive at their playing destination the day a tournament starts; they always arrive early to get comfortable with the course and environmental conditions. Ball flight characteristics affected by variations in gravity and air resistance are just another important adjustment in a long list, and usually professional golfers gain the required confidence for the conditions and make the appropriate adjustments after only a few hours of practice.

Another interesting aspect of gravity is terminal velocity—the limiting speed of an object in free fall determined by the ball's weight and diameter (diameter determines air resistance because it indicates the ball's surface area). Gravitational acceleration is the major factor contemplated when deciding whether to hit a low- or high-trajectory approach shot, and we know that in the absence of air resistance a released object falls at 9.8 meters per second after one second, 19.6 meters per second after two seconds, 29.4 meters per second after three seconds, and in a little over four seconds of free fall, the golf ball reaches its terminal velocity of 40.2 meters per second (90 mph). The small and fairly dense golf ball takes over twice as much time to reach its terminal velocity as a larger but less dense volleyball (35 mph), but reaches it over three times as quickly as a very dense 16-pound shot (325 mph).

Because it takes four seconds to reach terminal velocity, the golf ball always accelerates from the peak of its trajectory to the moment it lands, and this has strategic implications when deciding on what launch angle to use. For example, when a green is extremely hard, good golfers usually opt for a lower trajectory approach shot with more backspin. The alternative, the high-trajectory approach shot, can be disastrous if the ball hits the green at 90 mph and takes a high bounce that sends it bouncing off the green.

AERODYNAMICS

Aerodynamic forces are just as important as gravitational forces. It's useful to think of aerodynamic forces as altering the flight of a golf ball from what it would be in the absence of air (that is, in a vacuum). Besides being dependent on the local density of the air, wind velocity, and temperature, the effects of these forces are dependent upon the properties of the golf ball (shape, size, weight, texture of surface) and its velocity. Generally, the greater the velocity, the greater the effect of the aerodynamic forces, and since golf balls move both at a snail-like pace as well as velocities in excess of 150 mph, there is tremendous variability in aerodynamics.

Aerodynamic forces can be categorized as either lift or drag. Drag, which is always a retarding force, occurs in the direction opposite to the ball's flight (see Figure 4.1). While drag always shortens the ball's flight path by slowing it down, lift is more ambiguous and occurs in a direction perpendicular to drag. It can be either positive or negative, depending upon whether the lift is above or below the horizontal, respectively. Positive lift lengthens the flight path of the ball by opposing gravity, extending hang time, which means that the ball can move farther forward due to less resistance to its inertia (the tendency of an object in motion to stay in motion). Of course, there are exceptions, such as when a line-drive shot hits a low-resistance cart path surface. Achieving maximum distance across an asphalt parking lot would call for a very low trajectory since the ball's inertia is experiencing less resistance bouncing than it would covering the same distance airborne.

It's useful to think about the airborne characteristics of a golf ball by picturing a boundary layer of air near the surface of a ball as it moves. The air moves relative to the ball, so any object that moves through the air restricts air movement forward and to the side. Before coming in contact with the ball, the air is in an undis-

Figure 4.J:
The forces on a golf ball: The retarding drag force is always directly opposite the balls flight direction, and lift is always perpendicular to drag.

Figure 4.2:
The aerodynamics of both smooth and dimpled balls moving from right to left, (a) The slow movement of a smooth ball keeps the airflow pattern streamlined as it makes its way around the ball, (b) But once the smooth ball starts moving at a high velocity, separation of the boundary layer (turbulence) occurs, which creates a large low-pressure wake behind the ball, (c) Thus, dimples are used to trap and carry the turbulent layer of air farther along the surface when traveling at high speeds; the air "hugs" the ball, reducing drag, (d) Just as importantly, dimples entrap air along the surface to generate lift. As the ball spins from bottom to top, the air at the top of the ball, in the direction of the spinning motion, moves more rapidly than the air at the bottom of the ball, creating a downward force. According to Newtons Third Law, the air must, in turn, create an upward lifting force on the ball.

a) low-speed smooth ball

b) high-speed smooth ball

c) high-speed dimpled

turbed state and said to be streamlined. But once the air comes within a few inches of the ball, it can begin to break into a disturbed pattern referred to as turbulent flow. Figure 4.2(a) shows streamlined flow around a smooth, slow-moving ball. The slow movement of the ball, coupled with its smooth surface, keeps the airflow pattern streamlined as it makes its way around the ball. The flow past the ball is about the same as in front; there is no drag force since there is no momentum transfer to the air. As long as airflow is streamlined, Bernoulli's principle best explains the aerodynamics of objects in flight, stating that whenever airflow speeds up, a reduction in air pressure takes place. It's why when you blow a streamline of air between two helium balloons, the balloons move closer together, and it's why sailboats use a foresail (jib) to speed airflow past the leeward (that is, away from the wind) side of the mainsail. As long as the golf ball moves at a low speed, the flow remains streamlined. However, if the air's flow is turbulent, as is the case for a high-speed golf ball, Bernoulli's effect is negligible or nonexistent.

When a golf ball is traveling at high velocities and creating turbulent air, Bernoulli's principle is of limited use in explaining its aerodynamics. Figure 4.2(b) shows how the dynamics change drastically when the smooth ball moves at a high velocity. In the figure, the flow past the ball is now much different than in front, because the smooth surface of the fast-moving ball creates separation of the boundary layer (or turbulence), which creates a large low-pressure wake behind it. This indicates that the mass of air trailing the ball is experiencing a change of momentum resulting from the force created by the ball acting on it from the left. The equal and opposite reaction force (the force of the air on the ball) retarding the ball's progress—the drag force resulting from the separation of the flows—is to the right.

You could easily jump to the conclusion that turbulence is bad when, in fact, it depends on the object. Ironically, the reason why the dimpled golf ball surface is so effective is that it creates beneficial turbulence, trapping a thin layer of air along the surface of the ball. USGA testing has determined that under similar launch conditions a smooth ball will carry 120 yards compared to 260 yards for a dimpled ball, thanks to the beneficial turbulence created by the dimples. Unlike a ball, aircraft achieve the best results when the surface of the airplane generates the least amount of turbulent air: Streamlined air is desired because of the relatively long cylindrical shape of plane bodies. If aircraft were shaped like a ball, it would make sense to dimple the aircraft, too.

Dimples modify the airflow as it makes its way around the high-velocity ball. The speed and direction of airflow slows due to interaction with the ball surface because the golf ball exerts forces on the air contacting it, and the air reacts in turn by exerting equal and opposite forces—lift and drag—on the ball. A rippling effect takes place as well: The layer of air in contact with the surface of the ball tends to slow down the adjacent boundary layer, and that layer slows down the next layer, and so on. Figure 4.2(c) shows how the ball traps and carries the turbulent layer of air farther along the surface of a dimpled ball traveling at high speeds. Instead of the air separating from the ball as in 4.2(b), it "hugs" the ball. It is important to remember that dimples do not reduce the frontal drag pushing the ball back because the cross-sectional area of the ball is the same; the real benefit of dimples is to reduce the drag force by decreasing the amount of air having its momentum changed, reducing the size of the low-pressure wake behind the ball, which lowers the pressure difference between the front and back of the ball.

Backspin changes the dynamics once again. The force of the air on a dimpled golf ball with backspin provides lift, with the amount of lift dependent on the speed of the ball, the speed of the spin, and the kind of dimpling (including size, shape, and pattern). Figure 4.2(d) shows the flow of air around a dimpled ball hit with backspin. The turbulent boundary layer is moving along with the ball as it spins. As the ball spins from bottom to top, the ball entraps and confines air. Therefore, the air at the top of the ball, in the direction of the spinning motion, is moving more rapidly than the air at the bottom of the ball. The air along the bottom of the ball, in effect, is traveling against the wind while the air layer at the top of the ball is moving with it. As the backside low-pressure zone of the ball rotates over the ball surface and downward, the ball is creating a downward force on the air, and according to Newton's Third Law, the air must, in turn, create an upward lifting force on the ball. This lifting force, which works like an airplane's wing, alters the ball's trajectory in an upward direction, and the faster the spin rate, the greater the lifting force.

Although this type of relative motion of the air at the top and bottom of the ball and the lift it creates is often attributed to Bernoulli's principle, it actually is predicated more on the Magnus effect. Whenever a ball spins, the Magnus effect acts on it. The effect is subtle and very complex. For a dimpled golf ball, backspin results in a vertical deflection upward (lift) as the air at the top of the ball is traveling relatively faster through the atmosphere than the air at the bottom. Whereas spin about the horizontal axis creates lift, spin about a vertical axis will cause the ball to veer left or right—a sidespin responsible for draws (hooks) and fades (slices). And in 1947 physicist John Davies discovered that the Magnus effect, paradoxically, can generate negative lift too. A perfectly smooth golf ballj, spinning at the same number of revolutions per minute that created upward deflection for the dimpled ball, results in deflection in the opposite direction, namely, downward.

The natural upward motion of the golf swing and angled clubface greatly aid in imparting backspin and generating lift, and as might be expected, the greater the clubhead loft, the greater the rate of backspin. Yet even when using a driver with a 10-degree loft angle, the golf ball can rotate at over 4,000 revolutions per minute (rpm), a rate of rotation greater than that of the best curveballs in Major League Baseball. That's why the beneficial aerodynamic lift caused by the Magnus effect in golf is probably greater than that found in any other sport. In baseball, the Magnus effect deflects a good curveball about a foot and results in long fly balls carrying about 10 to 20 feet farther when bat contact is slightly under the ball, both of which pale in comparison to a golf ball. Assuming two golf shots are launched with identical velocity, one with backspin and the other without, the ball hit with backspin remains airborne as much as two or three seconds longer and can end up traveling 60 to 100 yards farther.

• What's the Major Benefit of Dimples: Lift or Less Drag?

THERE is no question that dimples significantly improve distance, yet there still is considerable debate in the scientific community as to whether lift or drag reduction is more responsible for the distance boost. There is also a question as to whether dimple design matters; specifically, does a dimple design exist that achieves more distance from lift and another that generates more distance from drag reduction? Of course, this debate is meaningless to most golfers. The only thing that matters is that dimples work, and in a big way. Under similar launch conditions, you can expect a dimpled ball to travel 260 yards compared to 120 yards for a nondimpled smooth ball.

INITIAL LAUNCH ANGLE FOR DISTANCE

When a playing partner follows us on the tee and launches a drive at an initial launch angle a few degrees higher and ends up with 40 yards more distance, we're apt to question whether his greater distance was because of a better launch angle. More specifically, would ratcheting up our launch angle a few degrees result in matching his distance? Since there are several interrelated factors that determine how far a golf ball will fly, bounce, and roll, a definitive answer is impossible.

To achieve the maximum range using the minimum force, physics theory tells us to launch a ball at a 45-degree angle when the launch point and the landing point are at the same elevation. In practice, however, other variables can dramatically alter the optimum launch angle downward. Let's first look at the shot put since the optimal launch angle has been extensively studied, and air resistance, aerodynamic forces, and wind conditions are minimal for the massive shot. The optimal angle is a little less than 45 degrees primarily because the release height is eight to nine feet in the air. Further, the shot-putter can more efficiently launch the shot in a horizontal rather than a vertical direction, and thus the optimum angle of release decreases with a loss of velocity. For example, for an athlete putting the shot at 45 feet per second, the optimum angle of release would be 42 degrees, but if release velocity falls to 40 feet per second, the optimum angle of release drops to 41 degrees. And if release velocity drops to 37 feet per second, the optimum angle of release declines to 40 degrees.

It would be convenient to determine an optimum launch angle for golf, but since many more interdependent factors are involved, one universal launch angle is not ascertainable. The wide range of swing mechanics, swing velocities, shaft lengths, shaft flexes, driver lofts, and ball types used result in a wide range of outcomes. The smallest alteration in any input can change the collision dynamics and aerodynamic forces, which could alter the optimal initial launch angle upward or downward.

The golf ball is a lift-generating wonder that extends hang time, and giving a ball with negligible drag extended hang time allows it to travel the farthest in the horizontal direction. Lift allows for the golf ball to be launched with a smaller vertical initial velocity and a greater horizontal initial velocity. So, unlike the lift-deprived shot put, the upper range of the optimum initial launch angle for a golf ball is less than half that for the shot put. In the book *Search for the Perfect Swing,* Scottish physicist Alastair Cochran calculated that a 20-degree angle achieves the maximum carry. However, this may not maximize total length since the ball will not bounce and roll as far after hitting the ground at a steeper angle (though 20 degrees probably would maximize total length if the course is rain-soaked and soft). Physicist Herman Erlichson followed with a computer model in the mid 1970s that determined the maximum total range, including both the carry and bounce, is achieved with a 16-degree initial launch angle. He also showed that anywhere from an 11- to 20-degree angle would achieve very similar total range results. Given the same initial velocity and spin rate, he found that the total distance difference between an 11-degree initial launch angle and a 16-degree initial launch angle is only a few feet. Although Erlichson's optimum launch angle for distance analysis is quite compelling, there is a major problem with his findings: the assumption of the same initial velocity and backspin rate. For any given swing velocity, different equipment results in different results. An 8-degree lofted driver will result in a higher initial velocity and lower backspin rate than a 12-degree lofted driver, and as long as the 8-degree lofted club results in an 11-degree or better initial launch angle, it will deliver greater overall distance than the 12-degree lofted driver. Further, whereas the optimal launch angle rises as the initial velocity for the shot put increases, the golf ball's optimal launch angle declines with greater initial velocity. The lower-lofted driver delivers a slightly higher initial velocity, which, in effect, lowers the optimal launch angle for maximum distance.

Ideally, then, you want to consistently and effectively achieve an 11-degree or greater initial launch angle with the lowest-lofted driver possible. The big hitters on the professional tour get away with using 7- or 8-degree lofted drivers because of their enormous swing velocity. Going from a 100 mph impact to one at 115 mph (executing an identical swing with the same club) not only increases the elastic energy potential of the collision, but also raises the trajectory. There is more velocity, a higher launch angle, and more backspin, which increases as well because, all else being equal, the higher the velocity of impact, the greater the backspin rate, as there's more time for the ball to roll up the clubface.

Since backspin is so beneficial, you might be wondering why all the equipment manufacturers promote low-spin balls and lower-lofted drivers. Most golfers are misinformed about the benefits of lower-lofted drivers: The loft is decreased not to limit the amount of backspin, but to improve the coefficient of restitution (CoR) during the clubhead-ball collision, a measure of the elasticity of a collision. Golf balls have a CoR of around 0.8 at 20 mph, which drops to about 0.6 at 100 mph; this means that a ball dropped from 10 feet will bounce eight feet in the air, and when dropped from 100 feet will bounce 60 feet.

Any momentum imparted to the ball is divided between giving the ball spin and initial velocity—it's split between a vertical and horizontal component. The backspin generated from momentum is a benefit, but hyperbackspin assists flight very little and is detrimental when coming at the expense of the momentum-determining ball velocity. More spin, which is the result of the club's loft, the way that it's swung, and the compression properties of the ball, means less of the clubhead momentum is applied to the collision for distance. What is needed is a proper balance between spin and initial velocity to achieve the maximum distance: The higher the velocity of the ball, the lower the rate of backspin needed to generate beneficial lift. For example, a ball traveling at 150 mph and at 2,500 rpm of backspin can generate greater lift than a ball traveling at 100 mph and 4,000 rpm of backspin.

Besides the CoR properties of the ball, the compression characteristics and aerodynamic properties of the dimple design are important determinates of distance. If you're using a highly compressive three-piece ball with an aggressive dimple design (for more lift potential), the optimal launch angle will be lower off the tee, in the 12- to 14-degree range. The ball will leave the clubhead at a lower launch angle and will have greater horizontal initial velocity, but its higher inherent backspin rate means that the ball will generate more lift than a similarly struck, less compressive two-piece ball. The higher compression balls might have a lower CoR but travel nearly as far because of its greater hang time, owing to its greater backspin rate and lift for any impact velocity.

Well, then, you're probably wondering what all this means for you when shopping for a driver. In short, you want a driver that is well suited for the ball you are using. For example, the more compressive two- and three-piece ball used by Tour golfers is very responsive—creating more than ample backspin—which allows most pros to match it with a lower-lofted driver so that more momentum goes toward ball velocity and less to creating backspin. On the other hand, the two-piece balls used by most recreational players compress less and thus are better suited for higher-lofted drivers. (Over the last two years almost all golfers have switched from a three-piece ball to a two-piece model.) If you are considering a switch to a lower-lofted driver, only do so when your ball is regularly taking off at the upper region of the optimum launch angle range, which is 17 to 20 degrees. Unless you can match the high swing velocities and contact accuracy of professional golfers, it doesn't make sense to use a lower-lofted driver. Good timing, squarely striking the ball, and accelerating throughout the downswing to maximize clubhead velocity is far more critical for distance. Regardless of the launch angle, assuming square contact, the greater your driver's velocity at impact, the greater the distance the ball will travel. Given a choice between spending time fine-tuning your initial launch angle or boosting your initial velocity, the latter is far more critical. For example, a drive that takes off at 10 percent less than the optimal angle would reduce the distance by less than three yards for a 200-yard drive, but a reduction of 10 percent in velocity could reduce the distance by 30 to 40 yards. Unless you are skying tee shots or scorching extremely low line drives, you shouldn't worry about your driver's launch angle.

LAUNCH ANGLES FOR PRECISION SHOT-MAKING

Although an optimal launch angle is not critical for distance off the tee, it certainly is a key component for shot making. Determining an optimal launch angle is just another one of those subtle skills that top professional golfers possess, a keen knack for judging the right combination of launch velocity and angle, spin rate and direction (understood here as the decision to impart pure backspin versus backspin with a sidespin component). There is never one "right" answer, and for any given launch velocity, there's a range of launch angles that will result in, for instance, a chip shot dropping; likewise, for any launch angle, there's a range of initial velocities that will also hole the chip shot. Consider basketball, another game where a ball is deposited into a hole. A shooter can swish a 15-footer with anything greater than a 46-degree launch angle, but to allow for the greatest margin of error, it's best to use a minimum-force shot that achieves the maximum range (because the minimum velocity increases the chances of a friendly carom off the backboard), which requires, as determined by physicist Arjun Tan, a launch angle of 51 degrees for a 15-foot overhand shot. Although anywhere from 30 to 89 degrees can result in a made basket, any arc higher or lower than 51 degrees requires more velocity to score.

Correspondingly, there's a range of launch angles that a golfer can utilize for a good approach shot. The major difference in selecting the right launch angle for golf is that the elevation of the lie in relation to that of the green is not constant (unlike basketball, in which the basket is always 10 feet high). Whenever the ball is lower or higher than the green, the ball is going to land, respectively, sooner or later in its flight path. Thus, you have to choose an iron one step shorter or longer for every 10 to 15 yards of depression or elevation.

There's also the gravitational acceleration factor to consider. The further the golf ball ascends, gravitational acceleration assures a high velocity by the time it reaches the ground—limiting your ability to control the bounce and roll. On a hard green, it's much easier to stop a ball when you're hitting up to a higher plane since there will be less gravitational acceleration, though when greens are soft, the difference in elevation between your lie and the green is a nonfactor. The bounce and roll are only marginally greater if the original lie is below the green, because the angle of descent is the same and the ball burrows into the soft surface on landing, which absorbs much of the additional gravitational acceleration.

Determining spin—whether backspin, sidespin, or none at all—is critical for all types of shot-making. Backspin is of the greatest significance because it slows the shot in flight by decreasing the air pressure above the ball and increasing it below. This alters the flight path so that the ball descends more vertically than one with little or no spin. Then, upon landing, whatever forward inertia the ball still possesses is countered by the backspin, which creates greater friction between ball and surface. In theory, backspin allows the ball to come to rest closer to where it lands. Thus, if you want to put the brakes on a landing, either hit the ball at a very high trajectory, with plenty of backspin to give it lift—altering the course in such a way that it falls close to perpendicular—or you can hit a lower trajectory shot with exaggerated backspin.

Selecting a trajectory is usually an issue only when a full swing isn't needed to reach the green. Generally our shots come down to working the clubface or just letting the loft of the club determine backspin and launch angle. The amount of backspin you impart to the ball is only partially determined by the velocity of your swing and the loft angle of the club, as soft, magical hand work is another invisible skill of professional golfers: Predictably, their hands and wrists can impart far more backspin with any iron than your average golfer. For example, PGA players sometimes decide to hit lower trajectory shots with exaggerated backspin, optimizing backspin at the expense of loft and velocity by using more of a "peeling" action as the clubface comes under the ball.

These lower-arc approach shots with exaggerated backspin are often favored since the flight path is shorter (making them more accurate), and the ball lands more softly (because less height means less gravitational acceleration), and there is more backspin to bring the ball to a stop nearer to where it lands.

Approach Shots: Farther is Better

Did you ever wonder why a professional golfer laid up 50 yards from the green when they could have just as easily knocked it to within 10 yards? As illogical as it may seem, many touring pros are more confident in their skill from greater distances, a phenomenon unique to golf. In basketball, skilled players shoot a higher percentage from 15 feet away than from 25 primarily because the lateral margin for error is larger with the shorter shot. The golfer is also trying to drop a ball in a hole, but the difference is that the intent of an approach shot is not to make it—though it is a much welcome result—but to get the ball in the best position to one-putt. To that end there are two things that the longer approach shot allows: lift and better bite. Lift alters the ball's trajectory so that it drops at a steeper angle, which allows for a more predictable bounce and roll when landing a ball on the green. At a distance of only 10 yards from the hole, the ball is hit too slowly to generate lift. The 50-yard approach shot, on the other hand, comes down at a much steeper angle, not just because of the more severe initial launch angle, but also because the lift alters the trajectory so that the ball drops at a steeper angle. Moreover, it's also easier to generate more backspin with a longer swing.

For some very low-velocity short chip shots, there is actually little or no lift. Backspin used for the short chip to the green can have a major impact on what happens upon landing, but has very little effect on the flight of the ball, however it's generally the case that anything greater than a half swing will generate a high enough backspin rate and velocity to create lift. In other words, any shot from greater than 40 or 50 yards out should have a

• Fliers—What Are They Anyway?

WHEN hitting from the rough, thick grass often gets sandwiched between the ball and the clubface. With little or no direct contact between the clubface and ball, the shot is hit with topspin, not the typical backspin. An errant swing where the base of the club strikes the top half of the ball also results in a topspin flight. Amateurs and professionals alike believe topspin shots have greater velocity off the club, fly faster and farther, and bounce a longer distance.

Called a "flier," it's actually a misnomer. Although fliers bounce farther because they hit the ground at a higher speed, they actually "fly" less, staying airborne for a shorter time and distance. Faster flight is an illusion; it appears to fly faster because of its lower and downward breaking trajectory (called negative lift), its higher speed when it hits the ground, and much faster and longer bounce and roll.

Off the club, there's no difference between the velocity of a ball with backspin and one with topspin, but in less than a second, the two balls start to take different trajectories as the flier experiences no lift or a negative lift and the shot with backspin rises dramatically. The negative lift, coupled with gravitational acceleration, results in a ball with topspin taking a more parabolic trajectory, while on the other hand, a typical backspin shot, because of its greater positive lift and less frontal drag, carries farther and drops more perpendicularly. The backspin ball actually flies farther, longer and bounces less, because air has less impact on the ball's inertia than does the surface of the fairway and rough that slow down the topspinning ball. Unless a topspin ball hits a very favorable surface, like a paved cart path, the result is that the sum distance of the carry and the bounce always will be less than that for the balls with backspin.

As inappropriate as the name flier seems, perhaps the name stuck merely because they don't appear to hang or float like typical golf shots.

sufficient amount of velocity and beneficial backspin to create lift that brings the ball down more vertically so that it sticks the green like a lawn dart. This is not to say that you cannot generate a significant amount of backspin with the 20-yard approach shot, but its much lower initial launch angle and decreased lift result in the ball hitting the ground at a flatter angle, which makes it much tougher to stop the ball near where it lands. If the golf ball was larger and lighter, lift could be generated at a much lower velocity, but for any given rate of rotation, the small, dense ball requires a velocity one-and-a-half to two times greater than a baseball and about two-and-a-half to four more than a volleyball to create Magnus effect deflection, or lift. These dynamics explain why professional golfers usually don't hit with maximum power on a 320-yard par-4. They avoid the higher probability of error associated with trying to cream a 300-yard drive and instead concentrate on a more accurate, smooth swing that will result in a 250-yard drive. From 70 yards out, they are in much better position to hit an approach with enough backspin for lift and bite after landing.

PLAYING IN WINDY CONDITIONS

Because it is small and lightweight, the golf ball is subjected to substantial wind effects while airborne that can alter where it lands by 40 yards or more. Every golfer knows that wind conditions can significantly increase scores; however, there is a misperception that higher scores are always due to a "power-robbing" effect, which isn't the case. In fact, from a cumulative 18-hole distance perspective, the wind can help as much as hinder, because for every hole against the wind, you're sure to eventually hit with it. If you understand the aerodynamics involved and vary your play accordingly, the wind can be a net benefit on a day with a light breeze (3,000 net yards in driving distance on a breezeless day can easily increase to 3,150 net yards when a 10 mph wind is blowing).

The real reason why wind pads scores is the increased difficulty of shot-making. On a very windy and dry difficult course, the average score for a round in a professional tournament could be 72, but in breezeless conditions following an overnight rain (making the greens soft and forgiving), the average score can easily drop to 67.

The physics of playing in the wind are quite simple. Whenever a golf ball is airborne, it will be subjected to lift and drag (resistance) forces, just like in breezeless conditions. What wind does, however, is change the magnitude of the lift and drag forces. For balls hit directly into the wind, the relative velocity of the air going past the ball naturally will be greater than if there was no breeze. Air velocity going past the ball will be equal to the velocity of the ball plus that of the wind. Lift and drag forces are determined by this total velocity, not just the velocity of the ball or that of the wind relative to the ground.

A ball hit into the wind will have greater lift because it's backspinning through air that is moving more quickly by, and has more drag because the ball's effective frontal area is creating greater resistance, explaining why a 120 mph drive hit into a 30 mph headwind experiences the equivalent drag and lift of a 150 mph drive in breezeless conditions. Hitting into a headwind will always result in greater lift. It's the same reason a hang glider always launches into the wind; with a meager forward velocity (small or no run-up), this is the only way the glider can use the wind's lifting force to rise into the air.

Despite the increased lift from hitting into the wind, the golf ball doesn't travel as far as it does in calmer conditions since the additional drag force is greater than the added lift. Conversely, less lift and drag occurs when the ball is hit with the wind, and it travels farther because it still has ample lift, and the very light ball encounters less drag impeding its flight. But for a 30 mph wind, the drive into a headwind will come up 83 yards shorter, but only add 25 yards for a tailwind.

To best take advantage of windy conditions, drives into a headwind should be fired at a lower trajectory and with less backspin, and vice versa for a tailwind. Maximizing tailwinds can be accomplished by either moving the ball forward, using a 3-wood, or carrying something like a 14-degree driver instead of a 4-wood on windy days. It's possible to attain more overall distance when the breeze is 10 mph or less, but for wind speeds over 15 mph, there always will be a greater overall loss, as the gained distance with a tailwind will be less than that lost when hitting into a headwind. Therefore, learning the headwind adjustment of using a lower trajectory is far more important than picking up the corresponding tailwind adjustment.

As far as shot-making is concerned, wind alters the flight path most when the ball has been struck with a higher trajectory, slower velocity, and lower rate of backspin. You have a far greater margin of error playing into a strong headwind than a strong tailwind, because the additional lift from hitting into the wind slows the ball more in flight, brings your approach shot down more vertically, and the backspin stops it much more quickly. It's also important to remember that no matter how hard the wind is blowing, the dimpled, backspinning golf ball not only generates lift that increases distance, but also helps maintain a truer flight path—that is, the actual trajectory better matches the intended one. Moreover, fades and draws will be exaggerated for balls hit at higher velocities into the wind (due to greater drag and lift) and will be less affected at lower velocities with a tailwind.

If winds blew only directly in front or behind the ball, adjustments could be made quite easily. But golf is not this simple because there rarely are direct headwinds or tailwinds, and crosswinds and 45-degree winds require additional adjustments. The ball is still subjected to drag along its path but also must contend with crosswind drag causing the ball to veer in the direction of the wind. And naturally, it will drift in the direction of the crosswind and will be subject to much greater veering the stronger the wind blows.

Another factor that plays havoc on shots are tree lines, which have the potential to both alter the direction and speed of the wind. According to the Bernoulli principle, when air moves through a tapered pipe, streamlined airflow constricts, increasing wind speed, while in the opposite direction, streamlined airflow moving apart decreases wind speed. Thus, heavily wooded courses can be very challenging because airflow constricts and speeds up while whistling through the canyons of trees. These canyons also change the wind's course; for example, if the wind is blowing from the west, and you're hitting the ball northwest up a tight tree-lined fairway, a 45-degree wind is more likely to play like a head wind. Keep in mind also that obstructions affecting wind conditions and ball flight are mostly those near the target, where ball velocity declines the most, and not where the ball is being hit. Flipping a few blades of grass into the air to gauge wind speed and direction is helpful, yet not nearly as critical as reading those same elements near the target by watching the flag and trees near the green.

Gusty conditions present the most difficult challenge, and when the wind is whipping and swirling in ever-changing directions, it becomes a guessing game to try to predict the conditions that will be encountered by the ball once it's airborne. As surprising as it may seem, however, some professional golfers look forward to windy days, and no, these golfers are not masochists. They feel a competitive advantage from conditions that call for another layer of analysis. Chaotic, gusty weather is more responsible than anything else for shaking the confidence of a professional golfer; while a few manage to thrive, the majority of them flounder. Those who struggle might be quite adept at reading a steady breeze, but fall apart when conditions turn ugly.

Playing in heavy wind can be extremely tricky because there's no such thing as a perfectly consistent breeze. The wind never blows steadily, nor does it always come from the same direction. Winds tend to blow in one direction but gust from a slightly different one. A steady headwind can become a gusting 45-degree wind or even a crosswind, and these unpredictable changes in wind direction turn a golf course that's difficult under normal circumstances into a nearly impossible one on blustery days. The 5-iron shot that would land within inches of the cup during the expected 10 mph headwind might blow 20 yards off-line if a 30 mph crosswind gust kicks up instead.

Obviously, then, the better your ability to predict gusts—their direction, strength, and duration—the better will be your results. Before playing, consult the wind forecasts that specify the speed and direction of gusts, and with this information in mind, before every stroke, consider which way the wind is gusting, the probability of it continuing, and how it will be affected by the tree lines. You should also keep a watchful eye on cloud formations for a clear indicator of gusts: Small cumulus clouds mean small wind gust changes in direction, speed, and frequency, and large, well-rounded formations of cumulus indicate the strongest breezes.

Variations in wind gusts can be attributed to convection (the transfer of heat by the movement of air masses), enhanced by the formation and dispersion of cumulus clouds. Convection creates the gusts along the surface of the earth because it brings the stronger winds blowing above downward. The cool, fast-moving air from above slows due to friction with other cells of air, and the wind near the surface is slowed and warmed by contact, for the ground is warmer than the overlying airstream. This creates wind gusts that circulate up and down, rolling along like tumbleweeds through a prairie. Although gusts occur everywhere, far greater convection takes place over dark surfaces such as plowed fields and asphalt parking lots because the sun heats up these surfaces more than lighter colored areas. Gusts are usually strongest in mid-to-late afternoon, after the sun has had a chance to heat up the dark surfaces. Professional golfers who draw late tee times are usually at a significant disadvantage on a bright sunny day in which the sky fills with large, well-rounded cumulus clouds. They are likely to experience wind conditions far worse than those golfers who started play earlier in the day.

Aside from figuring out how to make adjustments for breezes, there is also the matter of timing your swing for lulls. As a general rule, the life of a gust lasts for at least 700 to 1,000 feet downwind and can continue on for much longer. Thus, if hitting into a headwind, you can choose an optimal time to hit your shot by looking at the flag. As a 25 mph gust subsides and begins to lull the flag at 100 yards, it should completely blow by your position in ten seconds, but don't wait the full 10 seconds for the gust to completely subside before hitting the ball. The wind blows the ball off course mostly at the tail end of the ball's flight—near its peak and during descent—when both its velocity and backspin rotation rate are on the decline. So, hitting the ball 6 or 7 seconds after you see the flag beginning to lull minimizes the amount of unwanted gust-related veering by lessening the chance that the next gust will blow at full tilt on your ball's downward trajectory. The biggest mistake made by most golfers is to wait for the gust to subside where they are hitting the ball, which exposes them to the risk that the next one will start up with a vengeance while their ball is airborne near the green.

Bouncing, Rolling, and Skidding

One way to get out of the wind is to stay near the ground. To this point, we have looked at environmental factors solely for the airborne golf ball, but eventually gravity brings the golf ball to the ground, where considerable bouncing, sliding, and rolling take place on the way to the hole.

How well a ball moves across a surface depends on the frictional properties involved. When two objects slide across each other—as when a ball slides across the ground—a frictional force exists that retards movement. The coefficient of friction, which is the frictional force for a level surface relative to the weight (because the heavier the object, the deeper it sinks into the surface it's pressing against), depends on the nature of the surfaces in contact. For example, sandpaper on sandpaper has a coefficient of friction of 1 or more, a rubber tire on dry concrete about 0.8, and a hockey puck sliding on near frictionless ice around 0.08.

Friction impedes the progress of any object, but is also the reason a ball rolls; without it, a ball would slide like a hockey puck. Friction creates a braking force at the point the ball meets the surface, and this braking force creates torque, or rotational motion, which causes the top of the ball to move faster than the bottom, which sends it tumbling.

For a ball rolling on a smooth hard surface, friction isn't a factor because of the ball's extremely low rolling coefficient of friction (about 0.01). In other words, when a putter strikes a golf ball, it moves as easily as when a hockey stick strikes a puck; however, golf isn't played on a smooth hard surface. The ball travels across surfaces of ever changing contours with a range of friction coefficients. At one extreme are dry low-cut greens, the nearest thing to a smooth hard surface; at the other extreme is deep fringe, an impossible to calculate, varied surface.

Since inertia tends to keep the ball rolling on a green, golfers need to develop their "stopping" awareness and skills. Understanding backspin and the strategic use of gravity are the stopping skills necessary to strategically leave an uphill putt near the cup. Since gravity is an ally, you don't have to worry much about keeping the ball close to the hole if you miss; it will put the brakes on your ball so that it doesn't roll too far past the cup, so you can be aggressive. On the other hand, if you have a downhill putt, gravity is an enemy. You can't afford to strike the putt too firmly, and your only choice is a minimum-force, maximum-range putt—a slow roller that comes to a stop near the hole if you miss.

For short shots from the fringes of the green, there's a good reason why professional golfers rarely putt the ball. On a very nonuniform surface, there is no way to accurately predict where the ball will go because of the high coefficient of friction retarding the roll and the high probability of bad bounces. By the time the ball fights its way through to the truer rolling green, the ball has been subjected to terrain that changes its direction and speed, often in unintended and completely random ways. For these shots, usually the bump-and-run (or chip) shot is the best option. The objective is to just clear the unpredictable fringe of the green so that all the bouncing and rolling occurs on the much more predictable green.

While backspin can be a tremendous benefit, generating lift for distance and providing stopping action once the ball lands, the downside is the greater unpredictability of the bounce and roll. Depending on how deep the ball "drills" into the green's surface and the way the spinning action affects its way out of the crater it made, it may go straight back, stop on the spot, roll slightly forward, kick back left, or kick back right. Therefore, when looking for the truest bounce or roll, you want a minimum of backspin and skidding. Topspin creates the least friction or braking force, and the minimally lofted or nonlofted putter is the only club that generates it. But few golfers realize that even a putt at a distance of 15 feet usually entails a skidding action for the first foot, and the longer the putt, the greater the amount the ball skids.

To get a topspin roll as quickly as possible, you want to strike the ball from the middle on upward so that the ball leaves the putter with minimal skidding before assuming topspin. An additional benefit of this striking action is that it lowers the chances of scuffing the ground, which is the primary and perhaps most aggravating putting error of moderate- and high-handicap golfers. The dreadful result of a scuff has the ball coming up way short of the hole and usually off-line. Scuffing can be avoided by choking up on the grip about an inch and addressing the ball a little lower on the clubface. This should dramatically lower your chances of scuffing and only marginally increase the chances of a topping error.

• Regulating Green Speed and Limiting Scoring

ONE of the easiest ways for the PGA Tour directors to put the brakes on score deflation would be to slow down the greens. Score deflation is as much a function of faster, truer-rolling greens as it is of better drivers and balls. Sure, better equipment has made it easier to get the ball within 15 feet of the hole for a birdie, but it has also become too easy to consistently drop that 10- to 15-foot putt.

The typical green today has a standard deviation in lateral error of around 2.5 inches for a 10-foot putt, meaning that out of 100 putts struck identically from 10 to 15 feet away, the ball will drop only 66 percent of the time. Conversely, you can expect the ball to miss 33 percent of the time due to its randomly veering from irregularities in the green surface. (If you need a quick refresher on standard deviations, skip ahead to Chapter 7.) Since the ball will randomly veer much more on longer-cut, slower greens, the standard deviation is much greater, virtually eliminating the "sure bet "nature of 10- to 15-foot putts, and not just because of the more irregular surface: slower greens increase the probability of lateral errors because the ball needs to be struck more firmly on putts from any distance. And the more firmly the ball is struck, the greater the probability and magnitude of a lateral error.

Equipment innovation will continue to lower scores, yet slowing down the greens certainly could curtail the drop-off a bit.

PUTTING STRATEGY

The most prevalent scientific analysis of putting uses something referred to as the "hole zone." If the hole was covered, the hole zone refers to all the "made putt" stop points—at and beyond the hole—on a flat and level green, illustrated in Figure 4.3(a). If the putt is dead center, the ball has a leeway of about 5 feet beyond the hole for a green with a typical coefficient of friction, but this distance diminishes quickly as putts deviate from dead center. For putts of 10 feet or shorter, the hole zone becomes progressively broader since aim can be off by much more and still result in a made putt. That is because the angle of lateral error is greater the shorter the putt, as shown in Figure 4.3(b).

To minimize the number of putts to hole out, Werner and Greig came up with rules of thumb based on the hole zone concept. For a low-handicap golfer, the optimal aim distance should be 7 inches beyond the cup for 5-foot putts, increasing to 13 inches for 20-foot putts, but diminishing to 7 inches for a 40-foot putt. This is slightly different than the optimal aim distance for a high-handicap golfer (30 handicap), which is 12 inches beyond, 14 inches beyond, and 3 inches short of the hole, respectively.

The concepts of hole zone and optimal aim distance are interesting exercises of limited practical importance since putts rarely take place on flat and level greens, and golfers do not approach the questions of direction and velocity thinking about the range of made putt stop points beyond the hole. Furthermore, for putts that break (curve), any change in velocity is going to alter both where the putt will begin to break and how it will continue to break. The only real-world putting rule followed by all professional golfers is to never leave the ball short. When the ball comes up short, there's always the question of what could have been. Aside from this universal rule, putt velocity decisions basically come down to a matter of confidence; the greater your confidence in your directional precision, the firmer the ball should be struck. If lacking that confidence, you're better served by hitting a minimum-force, maximum-range putt—a ball struck with just enough force to get it to the cup.

A putt can drop through the 4.25-inch (108 mm) diameter hole traveling quickly, barely rolling, or at a wide range of speeds

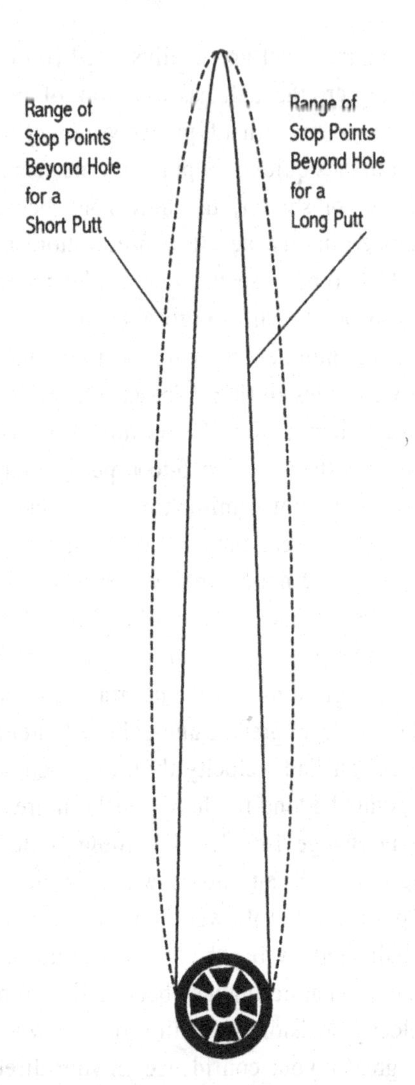

Figure 43(a):
An overhead view of the "hole zone" The hole zone is the area of the hole and beyond encompassing the entire range of made putt stop points if the hole was covered. On a typical level green, the ball could travel close to 5 *feet beyond the hole and still result in a made putt; any balls stopping outside this range would not drop. For short putts, the dotted line shows that the hole zone becomes wider since there is a larger margin for lateral error.*

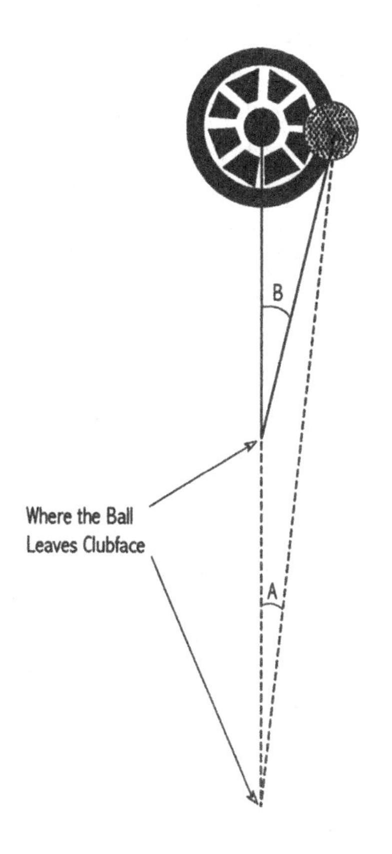

Figure 43(b):

An overhead view of the margin for lateral error for two putts of different lengths. The margin for lateral error for the short putt is greater than that for the long putt (angle B > angle A). That's why the closer you are to the hole, the more aggressively you can strike the hall, nearer to the higher range of velocities that will result in a made putt.

in between. Depending on the velocity chosen, the practical diameter of the cup can be functionally greater or smaller than what it appears; the faster the ball travels as it rolls up to the hole, the smaller the effective hole, and vice versa. Since the ball's diameter is a little less than one-third the diameter of the hole (the ball is 1.62 inches in diameter), a minimum-force putt provides the most generous margin for error, while at the other extreme, a ball struck very hard has a margin of lateral error that drops to zero. A high-velocity putt needs to roll near dead center of the cup to fall, and any putt with an angle to the left or right of dead center has a "smaller" backside of the hole for the ball to hit and drop. On the other hand, the minimum-force, maximum-range stroke results in the ball coming to rest at the cup, in effect, enlarging the target: The ball will drop if it's either 2.125 inches to the right or left of center (see Figure 4.3(b) again). Moreover, when putting across a sidehill, you always should target the upper edge of the cup so that there is the greatest chance of gravity helping drop the ball in the cup, because on the lower edge, gravity is more likely to cause the ball to rim out.

Reading the break of a sloped surface requires an understanding of the effect of gravity: Taking into account initial velocity and the perceived slope and speed of the green, how will gravity alter the direction of the putt? When will and how much will it break? For any putt that breaks, the minimum-force putt will curve the most on its way to the cup, increasing the probability of the ball breaking more than planned, since the slower the ball's speed, the lower its rotational inertia (which depends on the distribution of a ball's mass and the rate of rotation). Thus, the ball will not only naturally veer more, but is more likely to do so drastically if it encounters surface irregularities on its way to the hole.

Although the minimum-force, maximum-range strategy is usually best, it does not make sense from close range on difficult to read greens, especially if you're extremely confident of your accuracy and if the likelihood of a bad break or bounce is high. From medium to close range, most golfers are surer of their ability to accurately strike the ball and are more confident of their directional precision and therefore don't need the larger margin for error of a minimum-force putt. Striking the ball more firmly—what in golfing circles is referred to as more aggressively—increases rotational inertia, which removes some of the randomness of putting. If a shot is dead center accurate, there's a considerable margin for error in velocity. The ball can travel anywhere up to 4.6 feet per second (3.14 mph) and still drop (at 4.6 feet per second the ball will bounce off the backside of the cup, pop-up a bit, and then drop in). But this allowable error in velocity diminishes rapidly as directional precision (distance from dead center) wanes—reducing the area of the backside of the cup for the ball to hit and drop. Despite the advantages of aggressive putting, the penalty for missing is that the ball rolls well past the hole. That is why faster putts require far greater confidence in your accuracy; your objective has to change from getting the ball close to aggressively trying to make the putt.

Whenever the primary objective is to get the ball close, then again the minimum-force maximum-range strategy is preferred. Thafs why the objective is always to leave the approach shot at the lower elevation of the green so that you can putt uphill to the cup, which allows for a far greater range of initial ball velocities that will result in a made putt. And even when the putt is missed, gravity puts the brakes on the ball so that it doesn't scoot too far past the cup.

You often see a bewildered look from a pro after missing a putt, total disbelief as the ball either breaks too much or not enough. This occurs because reading a putt is a very inexact science. Even if you set up sophisticated survey equipment and took hours to precisely plumb-bob inch-by-inch the entire length of the putt, it does not ensure a made shot regardless of how well you aim; there are too many conflicting factors taking place that make determining the effect of an extremely difficult side-slope putt. Putts downhill across a slope, uphill across a slope, and ones where there is no uphill or downhill component will all break differently. That is because the forces on the ball will cause the trajectories to converge or diverge. Whether a golf ball rolling across a slope converges or diverges depends on how friction (the speed of the green) and gravity (which depends on the slope) affect the momentum of the ball, which has both linear and rotational parts and increases with the velocity. As a general rule, different ball trajectories diverge more when putting uphill and converge more when putting downhill. In other words, across the same 7 percent grade side slope and traveling at the same velocity, the uphill putt will break more than the downhill putt. To better understand this dynamic, picture streams of water at a fountain. The streams shooting out at a 45-degree angle above horizontal tend to diverge or break up; the streams shooting out at a 45-degree angle below horizontal tend to converge or stay streamlined. Unfortunately, these converging and diverging dynamics do not universally hold up in the real world because they rely on the assumption of the same ball velocity and coefficient of friction, which is far from what actually takes place: The ball velocity is always changing and the coefficient of friction will increase for increasing velocities. From any given distance, the uphill putt across a side slope is struck to achieve a much greater initial velocity and it comes to a stop much more quickly. Despite the converging trajectories, gravity can veer the slower-moving downhill putt more because the gravitational force pushing the ball off course has more time to affect the trajectory and there is less momentum to keep the ball moving in a straight line. The conflicting dynamics are maddeningly confusing, which is why putting can be such a humbling experience for even the most gifted professional.

Across-the-slope putts are particularly sensitive to errors in velocity. For a 10-foot putt across a 7 percent grade (given an average green speed), the trajectories will diverge for errors in velocity and converge for errors in direction; that is, your aim can be a littie off (as opposed to a level surface putt), but the initial velocity has to be extremely precise. A putt moving at 1 percent greater than optimal velocity (8.08 feet per second instead of 8.00 feet per second) will result in the ball rolling 1.2 inches above dead center, and a 1 percent less than optimal velocity (7.92 feet per second) will roll 1.3 inches below dead center. Compare this to a level surface putt: up to an 8 percent error in velocity will still result in a dead center shot, which is why correctly calculating the velocity for a side slope putt is far more important than for one on a level green; and also accounts for why it's extremely foolhardy to putt aggressively for downhill putts with a large side slope grade.

• Does the Effective Hole Enlarge for Side Hill Putts?

IN bowling the hook is highly coveted for a very important reason: A hookless bowling ball is only half as likely to result in a strike as a ball that enters the one-three pocket with 4 to 6 degrees of hook. It is intuitively appealing for golfers, who also happen to be avid bowlers, to think likewise about putts that break. The truth is that although the breaking golf ball does indeed marginally increase the effective size of the hole, it takes much greater precision in velocity to get the ball near the cup in the first place. As little as 1 percent greater or less than the optimum velocity for a 15-foot putt on a 7 percent side hill grade is likely to result in a miss, which compares very unfavorably to the 8 percent allowable velocity error for a 15-foot putt on a level surface.

Aside from determining how aggressively to putt from the contours of the green (either on an uphill, downhill or side hill), you also should consider the speed of the green. Whereas fast greens provide truer rolls and thus lend themselves to more aggressive putting, a minimum-force strategy for slow greens can be more advantageous since the effective hole enlarges by as much as a half inch, because when the ball is slowly rolling through the deeper grass near the edge of the cup, there are fewer blades of grass supporting the ball cup-side (because the groundskeeper shears them away while making the hole), and therefore the ball tends to drop toward the hole. If you picture the cup as having an inverted lip, the longer the blades of grass, the more inverted will be the lip of the cup allowing gravity to draw the ball more toward the hole. The big caveat, however, is getting close in the first place: If the ball gets close to the hole, it's more likely to go in, but the longer blades of grass are more likely to randomly veer the ball's path well before it ever gets near the cup's edge.

• Should You Be Trying to Read the Grass's Blade Lean?

BECAUSE of either mowing, the slope of the green, or the directional tracking of the sun, blades of grass lean. Popular folklore is that you can improve your putting by "reading the grain" and correcting for it. If you're in the majority of players who are not in the habit of lowering their eyes to check for blade lean, you can relax; you're not giving up a competitive advantage. No scientific research has been able to prove that leaning blades of grass veer putts. Those who are correcting are likely trying to fix something that isn't broken.

• The Human Factors

Shot-making skill is developed from years of practice building a feel and confidence for almost any imaginable shot. Professional golfers are excellent at running down a long list of factors before arriving at a shot-making decision, considering the lie, footing, incline of the original position and the landing, turf surrounding the lie and that of the potential landing, wetness of soil both at the address and at the target area, and then account for the numerous environmental conditions (gravity, wind, air density, and temperatures) that also vary significantly and require adjustments. Obviously, good golfers assess all these factors and have the confidence in their motor memory to make the necessary adjustments, intuitively visualizing how hard to hit the ball, calculating the trajectory, spin, landing, and the bounce and roll expected to best take advantage of the conditions. Nevertheless, what is so simple in principle is devilishly difficult in execution. Whenever the slightest bit of uncertainty, overfactoring, or overdiscounting sneaks into the picture, a shot that should have been well executed can turn into a nightmarishly bad one.

One of the major goals in the precision game is determining how best to use the loft and frictional properties of the clubface. Given the same velocity swing, how do highly skilled golfers hit a ball with a variety of trajectories and backspins? It's simply a matter of touch, which comes into play whenever a full power swing isn't necessary and is of foremost concern during the short game when targeting the green or pin. It's also a tougher skill to master than simply blasting the ball because it entails the combination of fine and gross motor control skills. Touch skills come from successfully drawing on your motor memory of thousands of previously executed swings—a wide range of velocities and ball spin objectives. When hitting a touch shot, there is less margin for error, more to go over during the preswing analysis, and a far greater chance that a choking doubt will foil execution.

In terms of the collision, you can improve your touch by either increasing dwell time (the period that the ball is in contact with the clubface) or maximizing the area that receives the force of impact (providing either greater ball deformation or more rotation up the clubface). Club loft selection, the arc of your swing, and what you do with your hands around the contact zone are the interdependent human factors that determine how and where the ball will travel.

Touch starts with "soft hands." Although the vernacular of golf commentators rarely includes this term, it's indeed a trait that the best golfers have in common with the top athletes in other sports. There are thousands of sensory receptors in the fingertips, far more than anywhere else in the body, and superior athletes have done the most to develop these receptors. Soft hands is a term usually used to describe bringing something traveling at a high-speed to rest or changing its direction. Football receivers bring a high-speed spiraling football to rest against their bodies, gradually decelerating its velocity by the skilled use of their arms, wrists, and hands; and tennis players execute drop volleys off 100 mph rocketlike ground strokes, absorbing most, but not all, of the impact of the shot so that the ball rebounds off the strings at an extremely low speed to carry a few feet over the net. Like the tennis racket, the golf club is an extension of the hands. The golfer and tennis player rely on soft hands to get a feel for the contact before the ball takes off at a launch velocity, initial angle, and spin rate (both back and side spin) to achieve the intended end. Skilled golfers don't just hit the ball; they toy with it, using the clubface like an apple peeler. This interplay of club and ball is entirely controlled by a feel for a soft touch action, and the more finely tuned the sensory receptors in your hands, the better your control.

The high coefficient of friction for wedges and short irons greatly expands the realm of possibility for precision shot-making. Of course, equipment design and materials are critical too; if the clubfaces of your short irons were a highly polished metal with a very low coefficient of friction, you could not hit low trajectory shots with high rates of backspin or sidespin, even with highly advanced touch skills. In fact, not much of any kind of shot-making would be possible; only the swing velocity and loft angle of your club would determine where the ball goes and what it does once it lands.

The length of the clubface also matters. When a professional needs to propel the ball a few yards, any iron can do the job, but their choice is the wedge and not the 2-iron for a very simple reason: the greater the surface of the clubface, the greater the possible backspin. With the wedge, there's a lot more clubface for the ball to roll up, and the wedge is used not for the loft that will send the ball into the air at a high trajectory, but for the area of the clubface imparting backspin. If you watch carefully, when a professional golfer hits a sand wedge, you may actually see the clubhead speed past the ball. The club moves under and past the ball for a split second before the ball carries upward and over the clubhead toward the target. In fact, double hits sometimes can occur when the clubhead comes up too quickly and the ball hits the clubhead as it rises.

Ironically, most moderate- to high-handicap golfers would be better served by using a 7- or 8-iron instead of a pitching or 60-degree wedge for shots from the fringes of the green. Since the clubhead velocity needed for a short chip shot with a wedge is about twice that with the 8-iron, it's much more likely that you will mishit (or not as cleanly strike) the ball. With the 7-iron, you sacrifice a small amount of backspin stopping action, but at this distance it's more than countered by the better directional and initial launch velocity precision afforded by the 7-iron. Moreover, less backspin means that the first bounce will be much truer, because of decreased friction on landing.

For professional golfers who are highly skilled with all their clubs, the decision is more complex. They usually will use a wedge because they are confident about getting under the ball so that it will softly land on the green and roll with less velocity. They can go for a specific target and get a more predictable response, particularly when the ball is landing on a downward slope to the pin. From the thick rough, the dynamics are a bit different. Pros go with the wedge when they feel a higher swing velocity will deliver a greater margin for error; swinging gingerly with the 7-iron heightens the chance of a very poor shot when the club is unexpectedly slowed fighting its way through the turf. In short, as long as a professional is not hitting out of deep rough or downhill to the cup—shots where there's little advantage from additional backspin stopping action—they too usually are better served by using a 7-iron and a shortened swing. But it all depends on the golfer: the greater the range of touch skills, the greater the options.

Touch shots directed at specific targets require a club-ball collision that is almost always less than full speed. This can be accomplished by either swinging at a slower velocity or shortening the swing length. The latter is almost always preferable since using the same swing acceleration mechanics for all your woods and irons makes it far easier to develop good touch. Think how much tougher it would be to develop a good feel if it was only achievable by making slight adjustments in swing velocity: How would your motor control distinguish the feel for an 88 mph impact versus a 90 mph collision? You couldn't; the nervous system simply isn't capable of delivering that kind of precision.

Shortening the swing is something that you can develop a feel for, and it keeps intact the mechanics of constant acceleration—good swing tempo—through the downswing. By using the same swing for all your clubs (a smooth acceleration throughout the downswing no matter the swing length), it's much easier to develop a repeatable fluid swing. Your motor memory is then left with the task of learning to associate the appropriate backswing length for any distance and club. If your 9-iron carries 120 yards with a full swing, and your 11 o'clock backswing carries 100 yards, you can easily develop the motor memory feel and confidence for the in-between backswing length that will carry 110 yards. In time, you will develop a greater appreciation and love for the challenges of the short game and actually look forward to challenging chip shots.

Hitting Fades and Draws

Fades and draws, or what otherwise can be called planned slices and hooks, are balls that leave the clubface with sidespin, which generates sideways Magnus effect veering. The reason for this curving applies also to the curveball in baseball: For each, the greater the rate of sidespin, the greater the amount of sideways veering, although, to be more accurate, deflection only occurs within a certain range of ball velocities. When the golf ball is traveling at a very slow velocity, regardless of the amount of sidespin, no deflection occurs. That's why you can't hit much of a fade with a wedge; short iron shots have a small window during the golf ball's time aloft when the veering can take place, probably between the halfway and three-quarter mark on its upward trajectory. Before then, the ball's velocity is too great, and as it begins to near the peak of its trajectory and during descent the ball's velocity is too slow to generate sideways Magnus effect deflection. The window of time also accounts for why a slicing tee shot will head straight up the fairway for the first 150 yards and then curve severely to the right after that. Unlike the short iron fade, it will start slicing slightly later because of its higher velocity, but will continue to curve through the peak of its trajectory and the descent. Gravitational acceleration slows the ball, yet for much lower launch angle tee shots, the horizontal component is much greater, and thus the ball's forward inertia is much greater. In other words, the ball's lower trajectory, which minimizes the slowing effect of gravitational acceleration, helps the ball retain a higher velocity throughout its flight, which results in greater overall deflection. Thus, the ability to hit effective fades and draws drops off considerably the higher the launch angle. If you need to hit a fade from 140 yards, you are better off using a three-quarter swing with a 5-iron than a 7- or 8-iron to achieve greater deflection.

When hitting fades and draws, what you're basically trying to accomplish is to slide the face of your club across the back of the ball—heel-to-toe for a fade, toe-to-heel for a draw. This can be done with either arm and hand action or by shifting your address position in relation to the ball. But you should not be frustrated if you are not equally proficient hitting a draw as a fade; for biomechanical and equipment reasons, it's much easier to hit a fade than a draw, explaining also why over 80 percent of beginners start out as slicers. As mentioned in Chapter 2, during the downswing the club's inertia flings the clubhead more outward, which triggers the eye-hand coordination adjustment to bring the club inward for contact; it's simply more natural to swing from outside to inside. Further, it's safer to move the ball across the clubface from heel-to-toe since the opposite is likely to move too far across and hit the base of the shaft.

Hitting Over Trees

Sooner or later every golfer is confronted with the problem of hitting over or around a tree obstructing a shot to the green. Usually flying the ball over is the desired choice since you can achieve the greatest distance, and it's possible to go directly at the hole. The big question is whether it's possible to clear the tree cleanly, and if so, what's the best club to use?

Conventional wisdom holds that the greater the loft, the higher the flight, but this isn't true. The 6-, 7-, or 8-iron sails higher than the pitching wedge or 9-iron because the ball leaves the 9-iron and pitching wedge at a lower launch velocity and much of the expected higher launch angle is sacrificed for creation of hyperbackspin. The loft of the wedge results in more backspin and a lower initial velocity and launch angle.

For any iron being swung, the peak of flight is reached in the 50 to 60 percent range of the total distance, nearer to 60 percent for faster swings and 50 percent for slower ones. Thus, if you're a fast swinger who hits a 7-iron 130 to 135 yards, your chances of clearing a tree are best when the obstruction is around 80 yards away, but as long as the tree is more than 60 yards away, you're still better off using a 7- or 8-iron than a 9-iron or pitching wedge. If you're a slower swinger who hits a 7-iron 100 yards, your chances of clearing a tree are best when it's 50 yards away, and the 7-iron is still your best choice as long as the tree is 40 or more yards away.

Difficult Lies

Since the ball isn't always sitting pretty, every golfer needs to develop a strategy for playing tight lies. For high-handicap players, higher lofted irons should be avoided since there's a very small margin for error up and down—the ball has to go through considerably more turf before ball contact. High-handicap golfers will experience fewer mishits using a three-quarter swing with a 6-iron than taking a full swing with a 9-iron. Low handicappers, who have considerably greater skill, improve their margin for error with higher lofted irons by playing the ball back at address, achieving more distance since it is nearly the same as using a club with less loft, and the clubhead has to travel a shorter distance through the grass (or water) before ball contact. Moderate-handicap golfers, who want to limit unnecessary variables to their shot-making, usually achieve the best results by addressing the ball normally with a less lofted club.

Putting

Champions almost always attribute a Tour victory to good putting. Likewise, many runner-ups sadly point to their putter as the reason they came up short. If a golfer cannot consistently drop difficult putts it's almost impossible to win. Post-tournament press conferences are peppered with the "ifs," "buts," "wouldas," and "couldas" about the regrettable one or two blown putts and the unfortunate "lip-outs" that separated the runner-up and victor.

In putting, the challenge is to make a "free" stroke to a specific target, "free" in that you have to hit the ball without trying to control it too much. Guiding, steering, or being careful with a putting stroke usually indicates doubt, either a lack of confidence in your ability or your read of the shot—determining the firmness of contact or the line that the ball should take. Many factors affect the decisions concerning the strength and direction of a putt, with the two most important interdependent factors being slope (either up or downhill) and speed (determined by moisture, softness, and length of grass).

An effective way to key in on a course's green conditions is to practice several different putting strokes. Vary your practice strokes by first taking swings that send the ball long of the hole, followed by a few that bring the ball up short. You can then focus your motor memory on the desired length, somewhere in the middle, knowing what impacts sent the ball long and short of the cup, and gaining greater confidence in an impact velocity somewhere between the extremes. You can lock in on the perfect speed since your brain now has the information and confidence it needs to determine how to strike the ball.

PUTTING STYLES

Whereas there are many similarities among professional golf swings using woods and irons, putting is different. No two golfers seem to putt exactly alike, with quite different stances, degrees of bending of the trunk, hunching of the shoulders, and placement of the hands on the grip. Among these various forms, it would be convenient to single out one style as superior to all others, but that's not possible. The wide range of putters in use does not help matters either: overall weight, its distribution (perimeter weighting), shaft length, and insert materials vary tremendously, more so than with any other piece of equipment in the golf bag. And obviously the best stance and swing mechanics for a standard putter will be different than for a long-shafted putter.

Although the difference in hand placements has been a rather subtle one in the past, a radical and intriguing method is quickly gaining popularity: cross-handed putting with the left hand below the right. Most right-handed golfers hold a putter with the right hand below the left, the same way they hold a driver or baseball bat; cross-handed putters position the left hand low, which proponents claim gives more stability and a better feel. The reason for this trend, as well as long-shafted putters, is that it helps make the putting motion more of a single pendulum swing rather than the double pendulum motion (around one axis in the mid-chest, another at the hands) used with irons and woods. Although a double pendulum swing is better for maximizing momentum, the single pendulum is superior for accuracy, and since accuracy is paramount, the cross-handed style makes it easier to more closely match intention with execution. The cross-handed method forces the left arm and putter to move together because they are more closely aligned in the same plane. Because the left hand generates the club velocity for the cross-handed putt, this eliminates any retarding, "feel-robbing," left-wrist torque (for a right-handed putter) during contact and follow-through that is common with the standard grip. The right hand is strictly for control, a fellow traveler with the left hand, guiding but not powering the swing.

The biomechanical superiority of the cross-handed style portends a slow migration as the next generation of golfers joins the Tour, and as top players experiment with ways to break out of putting slumps. Already many of the top professionals have resorted to this method, with Nick Faldo, Fred Couples, and Tom Kite among the most famous of the 40 or so who have taken up the cross-handed style. If not for the reluctance to give up years of ingrained motor memory practice, far more golfers probably would consider it as well.

Shot-making: Art or Science?

Getting the ball to go where you want is equal parts science and art. A professional golfer has a mastery of mind, body, and club that is akin to an artist's exquisite control of a paintbrush on a canvas, or a musician's deft fingertips on an instrument. When a pro fires a short approach shot, there is no easy way to judge whether the velocity, launch angle, and spin rate were optimal. We can only judge the results; if they're good, then we marvel at the shot-making ability that delivered that outcome. The art of shot-making is about considering the best range of scientific strategies and matching it with our best artistry as a shot-maker. There are a thousand different combinations of velocity, launch angle, and spin rate that will get the ball to the hole; the art of shot-making is choosing the combination that is best.

Although a professional golfer often has difficulty explaining the theoretical reason for a shot-making decision, he almost always has confidence that what he has in mind will work. Scientifically, these skills are developed through observation and experimentation, the culmination of the development of a wide range of ball-striking skills and a deeper understanding of how the environment determines where the shot goes.

5

Clubs and Balls: *Does It Matter What's in the Bag?*

MARKETING pitches for new golf equipment seem ever more imaginative. As hackers hook, slice, and swear their way around the links, they are easy prey for the game-improving promises from the manufacturers of the latest highly touted golf wares. Slickly produced commercials, backed up by endorsements from top pros, make true believers out of even the most skeptical player. After all, at the end of the day we all want to see a better scorecard and new equipment seems a good place to start.

If you remain one of the few unwavered by these marketing blitzes, you still usually face heavy-handed peer pressure. "Get with the times." "Don't be cheap." "These drivers really work; only a fool wouldn't use one." "Are you still using those antiques?" Paying big bucks for new clubs gives buyers an internal need to reconfirm their decision to upgrade by bringing more and more of the herd along with them. It becomes, in effect, a stampede, which starts because of the general outlook of most golfers. When given the option to improve themselves or their equipment, most choose the latter, which seems an ever so convenient shortcut to the promised land.

Unfortunately, trying to attribute improvements to any one factor, especially equipment, is a dubious exercise. The complexities of the golf swing make it extremely difficult to predict with any certainty how much new equipment will help, if at all. New gear always will have a different feel, often a very slight difference, but nonetheless it is a change. It's easy to attribute your better play to this different feel and ignore all other plausible factors. If you believe that you hit the ball better and farther or if you get a confidence lift, your scorecard is likely to look better even when the improvement is not attributable to the new clubs. Nevertheless, you naturally will credit your new equipment for the better play, a confidence born out of believing you have the latest and greatest technology. Confidence in your equipment is just as important as confidence in your analysis, motor memory, and visualization skills; you must be sure that your clubs will perform the way you think they should.

Golf equipment manufacturers relish the promotional opportunities that new designs and advances from the incorporation of space-age materials present. Both help bolster promises of better performance, and equally important, help distinguish equipment in a very competitive marketplace. Unfortunately, a golfer's perceived advantage is usually based on the ability of marketers to create expectations of better performance rather than on actually improved results. And although the manufacturer testing may prove that their new equipment is superior to other models, the tests are usually rigged with parameters more favorable to their design. For example, the claim of best driver is often dubious and inapplicable because the testing reflects neither real-world conditions nor your individual characteristics. Further, the development and testing of new equipment is still simply done by manufacturing prototypes and having pros try the equipment. Although often effective, the problem with such testing, and changes based on it, is that a golfer's swing is complex, difficult to measure, and varied, with no two swings exactly alike.

You can never know with any certainty whether the clubs and

balls you are using are the best match for your abilities. Nevertheless, to become a good judge of what equipment can and cannot do for your abilities, you have to first figure out what you do (how you swing), what you have the potential to do (the talent you possess to improve), and what the right equipment can do to help. Toward this end, you have to develop an understanding of the role of materials and designs. The two are intertwined because any change in materials can alter the performance of a design, and vice versa. Generally, golf club manufacturers seek stronger materials for durability, lighter materials that can improve swing velocity, and elastically "tuned" materials (i.e., the club and ball are tuned to each other) that improve the momentum transfer during the millisecond of contact.

Materials and Designs: Two Worlds Must Meet

From a scientific view, the collision between the club and ball entails an elastic energy collision, involving a combination of springlike and spongelike characteristics. When you think of springs, you probably picture a coil spring, but springs come in many forms and possess different characteristics. All materials have springlike qualities, and those materials found in your golf bag can act like a soft or stiff spring or something in between because their atoms and molecules are held together in a way that allows for movement back and forth. The law of conservation of energy states that energy can be neither created nor destroyed, although it may be changed from one form to another (e.g., it goes from kinetic to heat or sound energy). In the case of a collision between golf club and ball, a material can absorb energy or elastically return it, and in this instance, testing has found that stiffer springs are better for distance. You want a very stiff clubhead striking a very rigid ball.

Control and touch are another matter; softer clubface surfaces dampen energy and improve control because of the increased dwell time. Thus, since maximizing distance and improving control can be competing objectives, the best mix of materials for your driver are not going to be the same as those for your pitching wedge or putter. The goals are different, so the design and materials need to change to match the separate goals.

As few as 25 years ago, golf clubs were made of steel and wood. But today, the materials found in equipment are as varied and sophisticated as in any high-tech industry. Golf is just another sport affected by the wonders of exotic advanced materials. Substances such as aluminum, titanium, fiber-reinforced composites like fiberglass and graphite and ceramics have replaced wood and are in the process of phasing out steel. These new materials have wonderful properties, but unless you know why they were incorporated and how it's supposed to achieve its objective, you will never know if its introduction will help your game. Don't just buy into the marketing hype surrounding a new material; it may turn out to shave strokes from your score, but it's equally possible that it may be a nonfactor or actually hinder your game. Yes, the possibility of new gear having an adverse effect on your play is real. Golf equipment manufacturers select materials by what they believe will be the perception of the market as well as for actual performance concerns. Take titanium: As a word, *titanium* has a great sound that marketers love, much more tantalizing a word than good old *steel* After all, what can you say about steel anyway? It's well known, ubiquitous, and has been around for ages. In the case of a titanium clubhead, it is indeed superior to the wood and steel heads that it replaced: It has the advantage of being lightweight, more elastic (lowering energy loss), and strong, with a yield strength three to four times that of steel. And because of wider use and improved manufacturing methods, the cost is coming down too. Nevertheless, the advantage of titanium is a very marginal one, but because marketers can say the driver is made of space-age *titanium,* it's much easier for the potential buyer to justify purchasing a $1,000 driver. Aluminum is probably a superior material because it's lighter and only slightly less stiff or strong, but it would be far tougher to market a $1,000 aluminum clubhead that launches balls with that ugly pinging sound, most commonly heard around little league ballparks.

Clubs

Loft is the major determinate of trajectory. A smaller club loft angle translates to lower trajectory and spin, while greater loft angle results in higher and shorter ball flight with greater spin. Woods are designed primarily for distance and thus are only lofted at 6 to 15 degrees from vertical; irons are used for distance and accuracy and are lofted from 20 degrees for the 1-iron up to 45 degrees of inclination for the 9-iron; short irons and wedges are strictly for precision and are angled as much as 60 degrees; and putters are lofted from 0 to 3 degrees. As important as loft angle is in determining the trajectory and distance that the ball travels, there are four other important factors: the velocity of impact, the elasticity of the collision (from the materials and design of shaft and clubhead), the rotation that the clubface puts on the ball, and the amount of momentum that is transferred to the ball and not dampened within the club or body. Because there are four different variables, even the smallest change in swing mechanics, club design, or materials can have a dramatic effect on the results.

The multitude of tradeoffs involved makes the task facing equipment designers daunting. A change that might improve the momentum transfer at impact could also create too much additional backspin, resulting in a net loss in distance. Further, these alterations might affect a pro differently than a hacker, resulting in too much backspin for the pro, but just the right amount or too little for the hacker. Your "built-in" input errors will be compounded by a shaft or clubhead design that is not well suited to your swing; likewise, a design that is well matched to your swing will minimize the consequences of your mistakes. As manufacturers have developed a better understanding of the shaft and clubhead dynamics during the swing, they have vastly expanded their product lines to offer equipment better tuned to the needs of a wider range of golfers. The following sections explain these dynamics so that you too can develop a better understanding of how clubs work and, hopefully, determine which ones best match your swing.

THE SHAFT

Of all the parts of the golf club that affect playability, the shaft is perhaps the most underappreciated. It's conveniently ignored, since it's neither touched nor does it rip into the ball, simply serving the utilitarian function of linking the grip to the clubhead. Nevertheless, the shaft is critical for two reasons: It elastically bends and straightens during the swing, and any weight that club designers save in the shaft (especially toward the tip) improves your chances of increasing clubhead velocity.

The power contributed by the shaft's flexibility, which depends on its material properties, is quite variable and hard to quantify. Originally constructed from hickory wood, tapered, tubular steel shafts quickly became the material of choice by the 1930s and was the sole one in use up until the 1970s. Up until that time, graphite composite shafts were not considered durable or strong enough. But as manufacturers have developed innovative ways to refine and strengthen graphite composites, they have been making strong inroads over the last several years because of their lighter weight. At around 80 grams (0.20 pounds), these composites are considerably lighter than steel shafts (120 grams, or 0.26 pounds) and can be more easily manufactured within a range of flexibilities.

As mentioned, flexibility is a key element of shaft design. It has been vividly captured by high-speed photography, but can also be easily demonstrated by holding a club at the grip and very quickly wiggling it back and forth. That wiggling is in essence loading (storing additional energy) and unloading (releasing stored energy). At the transition from the backswing to the downswing, the greatest bending takes place, but contrary to conventional wisdom, the elastic energy is not released on the first straightening forward. Throughout the downswing, the shaft unloads, reloads, and then is in the process of unloading again through the contact zone. At the top of the backswing, the clubhead loads (this is caused from the body starting forward as the club is still going back) and quickly unloads as the downswing begins, catching up to the wrists. A little later in the downswing, the shaft again reloads behind the forward path of the wrists so that at contact the shaft unloads, straightening out. The elastic energy from the shaft flexing forward, the result of the centrifugal effect (outward flinging) of the clubhead accelerating from 0 to over 100 mph in a fraction of a second, can create a whipping action that sends the club into the ball with 8 to 9 percent more velocity. When well tuned to your swing, this could easily add 10 to 20 more yards to your drive.

It was intuitively appealing for club designers to believe that if the standard 3 degrees of flex can bestow 8 percent greater velocity, perhaps a more whippy shaft with 4 or 5 degrees of flex should be able to give an even greater elastic energy boost to the collision. Instead, testing showed that these shafts are likely to be going the wrong way at the wrong time (something that you never have to worry about with a very stiff shaft). Because it doesn't unload on contact, there's no additional power boost, and it has a negligible or even detrimental effect on distance.

Another worry when using an extra "whippy" shaft is the increase in launch angle. Provided that the same swing is used, a highly flexible shaft will usually send the golf ball airborne at a higher trajectory and with more spin than a standard steel tubular shaft. Selecting a lower-lofted club can compensate for some of this problem, but its loft angle always makes it more difficult to get the ball airborne and does nothing to negate the tendency of a whippy shaft to exaggerate errors created by a twisting force that opens up the clubface angle, because more flexible shafts also twist more than usual. Contact between the clubface and the ball will not be flush, and the result is likely a slice. If you're prone to slicing the ball 20 yards off course, your shot might go awry by 30 yards or more with a whippier shaft.

Although the steel tubular club is tuned for the swings of most golfers, advances in material science are making possible composite shafts that offer the best of both worlds: They're stiff as steel but far more lightweight. Another advantage comes when the lower 30 percent of the shaft is lighter than the rest of it because the greater mass distribution at the lower extremes of the shaft retards clubhead speed (the farther the mass is distributed down the shaft, the higher the moment of inertia, the force it takes to get the club in motion). Grip weight is of little concern for the same reason: The closer the mass is to your hands, the less important mass becomes. Therefore, one of the great advantages of a multimaterial composite like boron-reinforced graphite, as opposed to a single material like steel, is that it allows for more latitude in mass distribution and permits the shaft to be customized to meet particular stress requirements, most importantly, tension and compression stresses in bending and torsional stresses in twisting. Composite shafts are assembled in such a way so that boron fiber reinforces the strength of the graphite by aligning most of the boron-fiber reinforcement along the length. Composites designed in this way make it easier to work around the lightweight-strength tradeoff by regional strengthening—adding more reinforcing boron fibers over the lower third of the length where the material needs to be thinner.

Shaft length is another avenue of experimentation opened by these new high-stiffness, lightweight composites. In particular, by using ultrastiff graphite to reduce torque (twisting) and incorporating boron to add strength, a 52K-inch graphite shafted driver can be made to weigh less than a 43-inch steel-shafted driver. The combination of a longer shaft that's also lighter weight is certain to increase clubhead velocity, for the same reason that a hammer thrower can throw farther with a three-foot long cord than with one only two feet long.

Almost all professional golfers now swing lighter, longer shafted drivers. Most prefer a shaft length in the 46 to 48 inch range, much longer than the roughly 42-inch average of twenty years ago, but despite the increased length, the composite design allows the newer clubs to be much lighter than the drivers of yesteryear. Interestingly, Werner and Greig, the authors of *How Golf Clubs Really Work,* determined that the optimal length depends solely on swing velocity rather than size, strength, or sex, surmising that the higher the swing velocity, the longer will be the optimal shaft length. They also found that an 80-gram graphite shaft would add up to 5 more yards of drive distance over the 120-gram steel shaft. (For bigger and stronger golfers, the difference between graphite and steel probably would be less.)

Although the composite shaft is clearly a better choice than steel for drivers, there are two reasons why the same advantage cannot be extended to irons: First, the shafts are shorter, mitigating a swing velocity advantage, and second, steel is much better at resisting the torsional (twisting) forces on the shaft. Since irons are primarily hit through turf and not off a tee, the twisting force is only a concern with irons. To reduce the rotational force on the shaft from off-center contact (occurring mainly with graphite shafts but also steel), club manufacturers have introduced fat-tipped shafts. By making the lower third of the diameter approximately 35 percent greater than a standard club's, the shaft has greater stiffness to better resist both bending and twisting. These shafts bend about 50 percent less and are far more effective than conventional ones at reducing the twisting of the clubhead for off-center hits. The only real downside is the different feel of fat-tipped shafts—the greater stiffness results in a much stronger feeling for the contact absorbed by the hands.

Despite the technical superiority of big tip shafts and composite materials, the move toward these new designs promises to be a slow one among Tour players because any change in shaft flexibility essentially alters their feel. A shaft change has far greater potential to affect a player's touch than any other, and there's nothing a touring pro fears more than losing the touch that took years of practice to develop.

In reality, the feel of a good hit actually occurs well after the ball is on its way. It takes some time for the vibrations at the clubhead to be first transmitted up the shaft to the motor control receptors in the hands, and then that sensation has to be sent to the brain for interpretation. That is why it probably does not matter what the exact feel (and sound) of contact is; golfers will learn to like any type of vibration transmitted up the shaft once they are accustomed to matching it with a good hit. Eventually all pros will use big-tipped shafts because all the up-and-coming young players are growing up with and developing a feel for the superior design. But should the average recreational player spend the money now for big tip composite shafts? It depends; as improved manufacturing methods bring down the cost and improve durability, the benefit is increasingly worth the cost premium, especially for slow-swinging, high-handicap golfers who need all the help they can get in generating clubhead velocity and improving accuracy.

• Golf Is the Only Major Sport that Doesn't Favor the Big and the Strong

WHEREAS size matters in all the major sports, golf is different. Taller and stronger golfers do not gain a significant advantage by swinging longer-shafted drivers than their shorter counterparts. The optimum combination of shaft length and clubhead weight has little relation to the size, strength, and sex of the golfer. Instead it depends solely on a golfer's swing velocity.

CLUBHEADS

From the standpoint of materials, the clubhead is the most challenging area for equipment designers. To hit the ball farther, the goal is to devise a lightweight material that's strong, hard, and acts as a stiff spring—dampening minimal momentum upon contact with the ball. Once again, there are a number of trade-offs inherent in the selection of any material, with the two most critical being strength versus density and elasticity versus density.

Better designs and mixes of materials have opened up a number of avenues to address these trade-offs: (1) If a clubhead can be made with a lighter material, you can have a bigger clubhead that can be swung at the same velocity as the smaller ones they replaced; (2) if the clubhead is made of a stronger material, it can be made thinner, which raises the possibilities of larger and more forgiving clubheads; (3) and if you're able to use a combination of materials, clubheads can be designed to meet conflicting goals. An example would be a lighter and stiffer clubhead material for more momentum transfer combined with a different material that achieves greater surface friction to create backspin.

USGA rules do not specify the weight of clubheads (they're all approximately the same weight), so there's no rule preventing the addition of lead tape to increase its weight. If you can swing the heavier club at the same velocity as the lighter one, it will result in a greater momentum transfer. But weight is seldom added because it almost always reduces swing velocity, which is a more important determinate of how far the ball travels. Thus, the trend among club manufacturers is to continue reducing clubhead weight.

To discover the true effectiveness of new clubhead materials and designs, it would take complex testing, which would have to be designed for the varied characteristics of different types of players. Two golfers might swing their drivers at equal velocities, and their slices might carry the same distance off-line, but the cause of their errant shots might be different, which means that the self-correcting elements of a club design will have a different effect for each of them. Because of this type of complexity, there isn't a universal design that is best for all golfers. You have to understand the objective of a new design or material and then determine how it's supposed to work before you can decide whether it will work for you.

With this in mind, here are the pros and cons of the designs and materials used for clubheads:

Driver Loft Angle:

The loft and frictional qualities of the clubface are the major factors affecting flight trajectory. For the greatest carry and roll, there's an optimal loft and amount of backspin that you want to impart to the ball: Physics tells us that the maximum range for any given ball velocity occurs when the initial (takeoff) angle is somewhere between 11 and 20 degrees, and any takeoff angle greater or smaller will produce a shorter drive. To achieve this trajectory range, you can select a driver with a loft of only 6 degrees or one as high as 15 degrees. Although not universal, a roughly 10-degree loft is a good place to start for a beginner; you can experiment with lower loft angles once your launch angle starts rising above 18 degrees. However, driver loft should be a low priority. You should be far more concerned with squarely striking the ball. It is more important to concentrate on hitting the ball with pure backspin (with minimal or no sidespin), than to worry whether your shot's flight trajectory is a bit more or less than the optimal takeoff angle.

Perimeter Weighting and Expanding the Sweet Spot:

Oversized drivers did not come into existence solely for their "you can't miss with this one" appearance. There's no debating that this was the perception equipment makers wanted to promote, but their benefit is less about "can't miss" and more about making misses better by enlarging the sweet spot (the area where there's zero or minimal rotational force on the clubface) for greater accuracy. Contact left or right of the sweet spot causes the club to twist in your hands and results in the ball sailing off-line.

Within the sweet spot, there's a center of percussion, a center for maximum coefficient of restitution, and a center of gravity. What matters most to club designers is the center of percussion, the position on the grip where the initial shock to your hands is at a minimum. It's the pivot point of the grip where the motion due to translation (up and down) and rotation (left and right) cancel each other. Not a fixed location, the center of percussion depends on the mass distribution of the club and the position of the hands. For example, if you choke up a bit or assume a higher grip, the center of percussion will change as well. Thus, the objective of club designers is to minimize the rotational force for contacts away from the center of percussion, which increases the farther impact occurs away from the center. Because manufacturers have incorporated investment-casting technology (used for complex and high-precision shapes), they have much greater latitude in designing clubheads with larger sweet spots. With the weight of these larger-head "hollow" drivers concentrated around the perimeter (in the face, surrounding walls, sole, and crown), the designer is able to adjust the weight locally to increase the momentum transfer potential and to resist twisting from off-center contact away from the sweet spot.

Experimentation with perimeter weighting in golf followed from the success of the same practice in tennis rackets during the late 1970s. Manufacturers realized that the best way to expand the racket's sweet spot was to expand the perimeter: the larger the racket head and the more the mass distributed toward the heel and toe, the larger the sweet spot. However, the benefit in golf has been less dramatic than in tennis, because the surface area of the clubface is much smaller than the racket head, and the solid material clubhead is a far less elastic platform than the strings of a racket. Nevertheless, perimeter-weighting advances have been a huge benefit to high-handicap golfers who have difficulty striking the ball squarely and on-center, and all modern day perimeter-weighted drivers are up to four times as accurate as drivers from the 1970s and early '80s.

The merits of enlarged sweet spots for skilled players are less apparent. In fact, you will see very few PGA professionals using the very large oversized drivers because they are much more accurate swingers and thus do not need a huge sweet spot. Further, Tour players swing at a much higher velocity, so oversized drivers do more to slow down their swings, because drag does not increase proportionally with velocity. For example, a bicyclist going 15 mph has to more than double her power output to cruise at a 20 mph pace. Obviously, the clubhead and shaft have a much better aerodynamic profile than the bicycle and rider, but a similar relationship between velocity and drag exists in golf. Drag from larger-headed drivers swung at less than 100 mph will only knock off a yard or less of drive distance. However, for the 115-mph rips common on the PGA Tour, drag would subtract considerably more yardage. The lower swing velocities found among women and senior golfers account for why oversized drivers are much more prevalent on the LPGA and Senior PGA Tours, but perhaps the major reason why so few PGA golfers use the larger-headed drivers is the unacceptable level of drag that affects their swing feel, and not the actual drag loss itself (maybe 5 yards of lost distance).

There is also a different feel problem (of sound and vibration) for cavity-backed perimeter-weighted clubs. Responding to the fears of players having to adjust to a different feel, manufacturers have filled clubhead cavities with polymer foam to closer mimic the touch and sound of real wood drivers.

Lower Center of Gravity:

Almost all manufacturers now market clubs with lower centers of gravity. During the very brief time of impact, only the clubhead and the lower few inches of the shaft participate in the momentum transfer, and a lower center of gravity is another way of concentrating more of the mass at the bottom of the club. If clubs A and B are of equal mass, but club B has a lower center of gravity (more mass near the sole of the clubhead), greater momentum can be imparted to the ball as long as club B can be swung at the same speed as club A. Usually, establishing a lower center of gravity is accomplished by constructing the majority of the clubhead with light metals, such as titanium, and placing denser materials near the club's bottom in the form of a thick sole plate. Besides adding beneficial perimeter weighting, it also minimizes the effects of topping the ball, which is a highly appreciated feature for high-handicap golfers, and can help create greater lift with a lower-lofted club.

Low *Profile Woods:*

For drivers, the trend is toward faces that are longer from toe to heel. From research conducted on where off-center contact occurs, Werner and Greig suggest that a face that tilts slightly upward at the toe would significantly increase sweet spot contact for high- and moderate-handicap players, and somewhat for low-handicappers.

Fairway woods are also getting longer in the toe-heel direction. Like low center of gravity clubs, low profile fairway woods get more of the club to where it does the most good, which is a benefit to pros and hackers alike. There is still uncertainty as to whether low profile fairway woods make it easier to get the ball airborne, which is why most instructors still discourage beginners from experimenting with fairway woods.

Cavity Back Irons:

Perimeter-weighting concepts have also been applied to cavity back irons such that most of the weight is distributed at the heel and toe rather than evenly spread or concentrated directly behind the impact area as in more traditional irons. Cavity back iron designs have made it possible to construct larger-headed irons with expanded sweet spots that weigh less than the smaller iron designs they replaced. This means that you can increase your swing speed and, at the same time, improve your accuracy. This is an innovation that can be a significant help to the beginner, but for the advanced player its advantage is far less pronounced.

Draw Bias Clubs:

Draw bias drivers are essentially self-correcting clubs that help the high handicapper by countering their tendency to slice the ball. The slice-correcting drivers do indeed work, but it is best to avoid their use since they foster the temptation to forget about correcting the mechanical problem that caused the slice in the first place. This is more than golf purist propaganda; these drivers may correct your slice, but you will still be short-changed on distance. That is because the biased driver limits the sidespin, yet cannot do anything to create the coveted backspin that provides the loft and carry down the fairway. If you can't correct your swing mechanics, these clubs are for you. But you should use them knowing you can't improve beyond a certain level.

Mixed Materials and Clubface Inserts:

Clubface surface properties determine the amount of spin the ball is hit with, and spin control is vital for both distance and shotmaking.

Although the USGA probably did not realize it at the time, the dropping of a seemingly innocuous rule in 1992 opened up a new world of experimentation for club manufacturers. USGA Rule 4-1(c) stated that "If the basic structural material of the head and face of a club, other than a putter, is metal, no insert or attachment is permitted." Dropping this rule meant that the body of the clubhead could be chosen with one set of design and manufacturing criteria in mind (such as being lightweight, strong, and stiff), while the material at the face may be chosen for another reason, like the ability to increase the dwell time and backspin.

Figuring out the trade-offs of mixing materials is a daunting job for equipment manufacturers. For example, a change that might positively affect ball velocity off the tee might at the same time negatively impact the spin rate, and the net result could be a shorter drive. Further, the effect for the pro who swings at 120 mph may be different than for the hacker who swings 80 mph. The mixing of materials has given designers greater latitude in developing specialized clubs for particular shots.

Although maximum momentum transfer calls for the stiffest club and golf ball possible, the body of the clubhead is often made of a less rigid material because of other manufacturing or design considerations. To compensate for this, manufacturers use face inserts made of very stiff materials: A common mix for driver and fairway woods are steel heads with very stiff and hard inserts such as zirconia ceramics and graphite to improve drive distance. Besides being hard and rigid, these lightweight materials also move the center of gravity toward the back of the club and allow for greater perimeter weighting, which makes the driver more forgiving of mishits. For the driver, the low-friction clubface surface results in the point of departure being much closer to the point of contact; the ball neither slides up the face of the club as much, limiting backspin, nor does it slip to the left or right, curtailing the sidespin that creates hooks and slices.

For wedges, the objective is different. Better control and backspin are what counts, so semistiff, higher friction insert materials are used. The greater frictional properties of the clubfaces allow the ball to roll up or across the clubface, which makes it possible to do more with the ball in shot-making situations.

Springlike Effect Drivers:

The USGA is embroiled in a controversy over the legality of springlike properties in drivers, which work similarly to the way a tennis racket elastically accelerates a ball. However, there is a major difference between the strings of a tennis racket and the clubface of a driver: The clubface is a much suffer spring—much more similar to a leaf shock (or spring) on a truck than to a trampoline.

The USGA outlawed thin-faced clubheads in November 1998 by enforcing their 1984 rule against springlike effect drivers: "The material and construction of the face shall not have the effect at impact of a spring, or impact significantly more spin to the ball than a standard steel face, or have any other effect which would unduly influence the movement of the ball. A clubhead is deemed to violate Rule 4-1 (e) if a golf ball, when fired at the clubhead, demonstrates a velocity ratio (ball rebound/ball incoming velocity) which is greater than the baseline plus the allowable test tolerance." The USGA claims enforcing such a rule is well warranted; otherwise "advances in technology could soon overtake skill as the major factor in success." The USGA probably outlawed these drivers to put a halt to this line of innovation not so much for what they can do, but more out of fear that incremental improvements would ruin the integrity of the game. The Royal & Ancient Golf Club of St. Andrews was more suspect and did not go along with the USGA ban. They claimed "that these factors [the springlike properties] have not added enough distance to be detrimental." Independent testing seems to buttress the opinion of the Royal & Ancient Golf Club of St. Andrews. Through the 2001 model year, legal drivers have matched the distance achieved with the ultrathin-faced drivers. Yet it is almost certain that as refinements in thin-faced driver innovations continue, they will eventually deliver greater distance.

• The USGA's Legal Problem with the Spring Rule

WHEN the USGA came up with the 1984 rule to outlaw springlike effect drivers, they were asking for trouble. The rule states: "The material and construction of the face shall not have the effect at impact of a spring, or impact significantly more spin to the ball than a standard steel face, or have any other effect which would unduly influence the movement of the ball." This vague rule, either from ignorance or by design, gave the USGA tremendous latitude to be arbitrary and subjective since: (1) All impacts have some springlike effect. (2) What does "significantly more spin" mean (is it 5 rpm or 500 rpm more)? (3) What does "unduly influence the movement of the ball" refer to (less drag, more lift, a more elastic collision, or something else)? As equipment manufacturers take the USGA to court, it will be extremely difficult for the USGA attorneys to defend this very subjective rule.

The biggest problem with the USGA ruling is that the springlike effect is not unique to the outlawed drivers. In reality, since all materials have elastic properties, all clubheads have springlike characteristics. Manufacturers, with good justification, can claim that any standard is arbitrary. The USGA rulemakers probably never gave a thought to springlike effect drivers when they were all made of solid wood. Nevertheless, on contact the old woods would compress, storing energy, and then return to their original shape, releasing that stored energy. The wood was fairly consistent, so the spring energy was as well. However, once manufacturers began to market hollow core drivers with thin clubfaces, it was only a matter of time before questions about springlike properties would arise.

Switching from wood to stronger and harder materials like titanium improved the elasticity of the club-ball collision by around 5 percent, and the elasticity of the ultrathin clubheads boosts the elasticity even more. That's because the elasticity of the collision no longer is solely a matter of how the clubhead compresses and returns to its original shape, but also how the tension energy is stored as the head bends and returned when the material flexes back to its initial position. The evolution of hollow core drivers gave clubhead designers the latitude to change the elasticity of the collision by either changing the material or changing the clubface thickness.

Springlike effect drivers will be a benefit for some and detriment to others; it depends on your swing and ability to accurately hit the ball at the sweet spot. You have to consistently hit the sweet spot of a thin-faced driver; to do otherwise results in both a decreased springlike effect and an errant angle as the ball leaves the clubface. To visualize the greater error, consider what happens when you land off-center on a trampoline: it hurls you uncontrollably toward the opposite side of the trampoline rather than straight up in the air; much of the springlike effect benefit is negated. Besides the traditional springs you have to control (within your body, the club shaft, and the material of the clubface), ultrathin clubfaces incorporate an additional spring that does not respond well to slightly off-center contact. Another concern is swing velocity: to get any of the secondary spring action, your swing must be at least 105 mph, because a more elastic surface benefits from an extended contact zone and more deformation of the clubface on impact, both of which allow for more of the momentum developed during the backswing and downswing to be transferred to the ball. (The swing must be at least 105 mph because the secondary spring, or tension, energy won't be created short of that velocity.) A 105 mph-plus swing is only attainable by better golfers, which in part explains why USGA Senior Technical Director Dick Rugge said:

"Up until the last few years, metal woods did not add appreciable distance to a well-struck drive when compared to a wooden driver. They did add more consistency and made the less-than-perfect shot a little better. In this way, they helped the less-skilled player more than they helped the highly skilled player. More recently, however, the USGA has determined that some thin-faced titanium drivers are capable of producing significant additional distance through spring-like effect in a manner that is contrary to existing Rules. WeVe also found that spring-like effect generally helps those with the most skill who need the least assistance."

The USGA claims that since the introduction of titanium drivers with a springlike effect in 1994, driving distance on the PGA Tour increased by two yards per year (through 1999), which is far greater than the one foot per year average from 1968 to 1994. Of course, a strong case can be made that other equipment innovation, as well as the much greater emphasis placed on strength and flexibility training in the 1990s, are equally responsible.

Regrettably the USGA has only itself to blame for this mess. Instead of setting a difficult-to-measure and controversial standard for springlike effect, the USGA could have devised a simple laboratory test to determine a minimal clubface thickness standard that would have made any secondary elasticity impossible. It just would have to be thick enough to prevent the material from bending to store and return tension energy. Since the span of the clubface is relatively small (with limited ability to bend), and the ball is extremely light and hit from a stationary position, a minimal thinness standard could still be very thin (so as to not add much to the weight of a clubhead).

Putters

Just as there is a large group of golfers who are always looking for the latest and greatest in new equipment, at the other extreme

• Big-Headed Drivers and the Confidence Compounding Effect

STUDIES have shown that modern drivers have added 10 to 15 yards of distance for professional golfers compared to those from the early '90s, and over 30 yards more than a 1980 vintage driver. But equally important is the fact that modern drivers keep on getting better at minimizing the errors caused by off-center contact. Therefore, pros not only are getting better driving accuracy, but also greater confidence to swing more recklessly, generating more clubhead velocity since the penalty associated with being slightly off-center at contact has been diminshed. Conceivably then, pros who regularly hit 250-yard drives in 1980 are hitting 300-yard drives today, with 20 of those additional 50 yards due to the heightened confidence to really go after it.

are the nostalgics who cling to the belief that the time-tested and well-worn vintage equipment of yesterday holds special magic. Although this belief is still held by many, testing and use have proven that today's equipment materials and designs are vastly superior to that of yesteryear. Touch is indeed important, and the putter is the foremost club of which a "well-worn feel" is an important characteristic, but the superior feel of new putters achieved by better designs and the mixing of materials to achieve seemingly conflicting goals more than offsets the psychological benefits of a vintage putter.

Because control, and not maximizing distance, is the objective, the materials selected and how they're incorporated are unique to putters. Whereas the insert for drivers are extremely stiff, hard metals and ceramics, those for putters are dampening materials that prolong the dwell time to improve feel. However, this better dwell time must not create backspin that causes the ball to skid. Establishing a longer dwell time and maintaining low friction are conflicting goals, which presents a real challenge for club designers. To improve touch, "soft" energy-dampening face insert materials such as thermoplastic polymers are favored to better "hug" the ball, resulting in better feel and limited skidding. Some of these materials appear to increase dwell time by a factor of about two!

When it comes to accuracy the putter is the one club where it really does not matter what modern, technically advanced model you use. Werner and Greig found that even when popular models are technically inferior, they still do not perform worse in testing. In other words, you should be highly skeptical of any putter that claims to reduce your lateral errors. What matters most is comfort and confidence, and whatever putter you feel the most comfortable and confident with is always the best choice. The only way to correct your lateral errors is to refine your mechanics. A new putter may increase the dwell time by a factor of two, yet that only ensures that you'll have a better feel for your lateral error.

BUYING CLUBS

Because manufacturers offer so many models, with several different material combinations, the shopping experience can be maddeningly confusing. Therefore, to select clubs that best match your swing, the following are some things to keep in mind:

1. Never buy equipment without trying it out first. It's not only necessary to test the equipment to see if it performs as advertised, but also to decide whether the equipment has the right feel for you. Feel is something that comes from developing a familiarity with your equipment, and if the clubs you're testing play considerably different than your present set, you have to consider whether you can adapt your sense of touch to the new gear. Even when you find equipment that offers better performance, there's the "feel learning curve" to remember: It's going to take time for you to become comfortable with the new clubs and develop the

same touch and confidence that you had with the old set. Though graphite composite shafts have many advantages, the reason only a few touring professionals are presently using them is because their distribution of mass and flexibility is not what they're accustomed to, which makes changing to the new shafts a gamble. Often the allure of big promotional money entices the top touring pros to gamble anyway, and usually they end up regretting it. Johnny Miller, once heralded as the next Nicklaus in the 1970s, switched equipment and saw his game plummet, never to return to its former glory, and he believes that the same fate recently befell Corey Pavin and Lee Janzen. Even when switching to technically superior equipment, count on it taking at least a year before your shot-making touch returns. Since an equipment change has major ramifications on a pro's ability to earn a living, they have a reason to worry about the impact on their touch. For the rest of us, it's a much smaller stress. We not only have less touch, but the little bit we have is far less ingrained in our motor memories.

2. Always take advantage of demo equipment. Compare the equipment to what you are already using as well as other new clubs you demo. And don't just hit the ball from level, perfect lies at a range; you need to try clubs from sidehill and difficult lies and from real life situations to see how they perform. This is particularly important in selecting irons, because the way the clubhead itself penetrates through the grass is likely to be different as well. Likewise you will need to adjust to a different shaft flex feel, an adjustment that's even greater when hitting out of deep grass.

3. Remember you are not Tiger Woods and neither do you swing like him. The best club for a superstar is not necessarily the best club for you. Tiger uses a 7.5-degree loft driver, and most recreational players would hit pond-skipping line drives with such a low-lofted club. In fact, if you're a high-handicap golfer, using the clubs of your favorite pro is highly likely to do more harm

than good. You have a unique swing, so you want clubs that match it. If you are unsure of what your strengths and weaknesses are, seek out the guidance of a professional who can evaluate your swing. To select clubs that best match your needs, you should make an honest assessment of your swing velocity and accuracy at contact.

- Heavy Hitters: Stick with a steel tubular shaft. If you're a big strong guy, an ounce or two in shaft weight is not going to make any difference in your swing velocity. If you also swing at over 105 mph, you might benefit from a springlike effect driver.
- Slow Swingers: You may want to experiment with a composite shaft driver. They'll help speed up your swing, and more of the mass will be at the end of the club where you want it.
- Wild Things (Inaccurate at Contact): The large-headed drivers and perimeter-weighted irons will both help expand the sweet spot, eliminating or minimizing the mishit torque that forces the club to twist in your hand at contact and sends the ball far off course.
- Straight Shooters (Accurate at Contact): You get the least value from the big-headed drivers. You don't need the big sweet spot, and a big-headed driver might even be a detriment if it slows down your swing velocity. You do, however, benefit from a lighter shaft and a lower center of gravity clubhead that gets more of the mass at the bottom of the clubhead where it can do the most good.
- Severe Slicers: You will profit from the so-called self-correcting clubs, but the advantage comes at some cost: You will feel less of a need to correct the flaws in your mechanics that cause the slice in the first place. You will never be able to develop the type of contact that creates the coveted backspin to improve drive distance nor learn to impart sidespin on command to hit draws and fades.
- Hard Workers: If you don't mind spending a lot of money,

you may want to take the "work my way up the ladder" approach. Start with the clubs that are most forgiving and offer some self-correction built in, and then as you advance, keep upgrading to gear that best matches your improving game.

Balls

Annual spending on golf balls has surpassed $650 million in the U.S. and $1.5 billion worldwide. As the number of golfers and courses continue to grow, and due to the depletability of the commodity in question, this very lucrative, high profit margin business is destined to continue growing. Neither the size nor the growth of this market has gone unnoticed by equipment manufacturers. Callaway, TaylorMade, and Nike are new to the golf ball game, ready to aggressively jockey for a share of the market dominated by industry titans Spalding and Titleist. The competition is fierce, and over the last few years, each of these companies has probably produced several thousand prototypes for evaluation, and competitive pressures ensure that this pace of innovation is only likely to continue.

The modern golf ball consists of either two or three parts. All balls have a core, and although the materials used may vary by manufacturer, all cores consist of variations of rubber compound concoctions. The core may be in the center of the ball, but it's not central to performance. What counts most are the materials and design of the one or two layers surrounding the center or, for very large core balls, the outer layers of the core. Of all the different types of balls sold, it has long been believed that three-piece balata-covered ones are best. In addition to a core consisting of rubber compounds, the three-piece ball is sometimes called a wound ball because of a spool-like middle layer that looks like very thin rubber bands woven together. The two-piece balls—consisting of a highly compressible acrylate or resin core and a very durable Surlyn (a patented thermoplastic resin) cover—are more durable and hard, resulting in greater carry. Up until recently, professionals preferred the wound balls for their greater feel, which is predicated on possessing the confidence to control the trajectory—both in the air and on the ground. Better control also means the ball is less affected by the vagaries of the wind and more responsive to draws and fades.

There is also the matter of feel and control at collision. Usually when professional golfers make anecdotal comments about better touch at contact, they are talking about ball compression, which means that the ball experiences greater flattening out, or deformation, at contact (because more of the surface of the ball is in contact with the club), prolonging the contact between the clubhead and ball.

Although three-piece balls were judged superior at the professional level, they were also far more expensive to manufacture. But thanks to advances in material science, the playability gap between two- and three-piece performance balls continues to narrow. Almost all ball manufacturers have developed two-piece balls that match the feel and control of a wound ball, yet still deliver the greater distance of the traditional two-piece model. They're doing this by enlarging the core and manufacturing thinner, tougher, and more responsive covers. These enlarged cores again use mixes of materials to achieve competing objectives, and while the innermost part of the core is a very hard, stiff, rubber material, the outer layers are progressively softer to improve the compression for feel.

One manufacturer advertises a ball as being both soft and resilient, or soft and elastic—a seemingly irreconcilable contradiction. The claim amounts to saying that the ball is both a spring and a sponge: soft enough to prolong contact for better accuracy, resilient enough to elastically spring back for greater distance. If golf ball cores consisted of only one material, this would be an impossible claim, but because they are made from many different materials and combined in many layers, and manufacturers keep getting better at selecting mixes of materials, these claims are actually becoming more and more accurate. In particular, two-piece balls keep on getting better at delivering both improved distance and control (compressive feel) and have thus finally begun to dominate the professional golf market. By the 2001 PGA season, well over 70 percent of players were using two-piece balls.

DISTANCE AND PLAYABILITY

The distance that a golf ball travels is primarily a function of its velocity and the club's loft angle, but is also related to wind conditions, the dimple pattern (determining the amount of drag), and speed of rotation (affecting the amount of lift).

The golf ball wars are largely about whose ball gets more momentum transfer during the collision with the clubhead. If all the momentum were transferred, 500-yard drives would be the norm. Of course, neither the ball nor the clubhead are perfectly elastic, and because of the short time span of the collision and the large difference in mass between the ball and the club, equipment manufacturers can never hope for a perfectly elastic collision. During the millisecond of contact, the clubhead only slows from roughly 110 mph to 100 mph; at that speed, the club still carries considerable momentum that has no chance of being transferred to the ball. Some momentum always will be dampened within the clubhead and shaft, some within the body, and the majority dissipates either within the ball or is "wasted," remaining with the club and your body during deceleration (the follow-through).

The coefficient of restitution (or CoR for short) is the ratio of the relative velocity (momentum) after direct impact to that before it. It's a measure of the elastic nature of a ball, and can vary from 0 to 1 (a superball comes close to 1 and a ball of clay would be 0). For a golf ball, it varies approximately from about 0.8 at 20 mph to 0.6 at 100 mph. According to collision theory, a perfectly elastic ball (CoR = 1.00) will leave the clubhead at twice the velocity of the clubhead's velocity just prior to impact. However, because the collision is not perfectly elastic, and all clubs have an

• When Is it Time for a Ball Change?

YOU'RE having a nightmarish round and are losing confidence with every stroke. You have already lost 10 new balls and feel there is no reason to gamble on losing another. So as you approach the long par-4 hole that requires a 200-yard drive to clear the pond, do you pull out another new ball or that battered old range ball? A sports psychologist might say the range ball choice shows a lack of confidence and is in effect a self-fulfilling prophesy: If you do not believe in your ability to clear the pond, hedging your bet by using the beat-up ball will not help your psyche at all. Physics is not on your side either: Your battered and deformed range ball might travel 20 yards shorter than a new ball.

For consistency of touch, it only makes sense to play the same ball all the time. But how about replacing balls after extensive play? In other words, should you keep battering around a ball until you lose it or does periodic replacement make more sense? It depends. Obviously you do not want to handicap yourself by playing with a deformed ball. A misshapen ball creates a greater cross-sectional area, which heightens the coefficient of drag, increasing air resistance. This increase in the coefficient of drag, along with more erratic airflow around the wobbling and spinning ball, reduces distance and adversely affects accuracy. An equally important concern is a decline in a ball's coefficient of restitution. This decline occurs not only from the pounding the ball takes during play, but also naturally over time. That is because in storage golf balls absorb moisture from the atmosphere that results in a loss of about 1 percent of elasticity a year, which translates into about 5 yards of drive distance for an advanced player.

Luckily new balls are far more resilient than those from 20 or 30 years ago. They can take a much greater pounding before assuming an out-of-round shape or losing significant elasticity from use and storage. Nevertheless, it's a good idea to periodically check the coefficient of restitution of the ball you're playing with by performing a simple test, bouncing it alongside a new ball.

When a ball starts to lose elasticity or starts to assume a less-than-round shape, throw it in the practice ball bucket. You neither want to play with a ball showing signs of fatigue in the "springs" holding the atoms in the ball together nor with one handicapped with greater drag from an out-of-round shape.

angle of loft, the factor is considerably less, about 1.46 for a driver, 1.30 for a 5-iron, and 1.12 for a 9-iron.

Between the club and ball, almost all of the momentum loss occurs in the ball. Since a golf ball is extremely hard to the touch, it is difficult to believe that it deforms significantly during the collision, but at contact the ball assumes an extremely out-of-round shape. That's because the atoms that make up this core are held together as if by springs that allow for movement back and forth, and the deformation happens in a millisecond, far too fast an event for even the keenest human eye to observe. Clubhead-ball contact compresses the springs; then as the ball explodes into the air, these atoms spring back to their original position, and the ball returns to its original size and shape. In the process some momentum is elastically returned and some is dampened within the ball. Many believe that the ball almost instantly reassumes its original shape, but in fact, the ball can deform from a multitude of hits or even a single blow and can remain misshapen for weeks. And if deformation from extensive play permanently strains the ball's materials, its flight path can actually become erratic. (See box.)

The march toward better and better balls is inevitable; the constant threat of competition and a better understanding of material properties promise more durability, explosiveness, and better playability. But the distance that a ball travels is as much about the aerodynamic properties as it is about the elastic properties. In particular, all dimples are not all created equal.

The story of the dimpled golf ball begins at the turn of the twentieth century. When the first hard-surface gutta-percha, or "guttie," ball was used in 1901, golfers began to notice that the longer they played with a particular ball, the more nicked and cut it became, and the farther it flew. This led to a series of experimental designs that brought over 200 different covers to the market between 1900 and 1920. The first experimental surface of note was the brambled ball, sometimes called the pimpled ball, which was indeed more effective than the smooth ball. But the brambled ball quickly became extinct after dimples appeared in 1908. This was not due to aerodynamic superiority, but primarily because dirt did not cling as much to the dimpled surface. Once the dimpled ball replaced the brambled model, innovations pertaining to the surface of the golf ball ended for a while.

It took until the early 1970s for manufacturers to discover that the size, depth, and number of dimples could have a dramatic effect on the distance and trajectory of a golf ball. Up until then, they were obsessed with distance and durability and devoted little, if any, attention to experimenting with dimple aerodynamics. This abruptly changed when Fred Holmstrom and Daniel Nepala, two San Jose State University professors, discovered that a better design would incorporate an asymmetrical dimple pattern, which would limit the hook and slice for the majority of shots that come off the club with sidespin.

The actual aerodynamics of the golf ball are quite complicated when explained in scientific terms like boundary layers, Magnus effect, and the Bernoulli principle, but the basic principles are simple: A spinning dimpled ball changes the air flow around the ball, which alters the flight path. Perfect backspin creates lift, which carries the ball farther, but if the ball comes off your clubface with sidespin, sideways deflection occurs, and the ball will hook or slice off course.

The reason why the dimpled surface is so effective is that it makes the air passing around the ball turbulent, trapping a layer of air that spins along with it. Called the Magnus effect, it creates a "beneficial" turbulence that decreases drag, and when hit with backspin, increases lift. A dimpled golf ball that carries and rolls 260 yards would reach only 120 yards without dimples. Backspin produces lift that alters the trajectory of the golf ball in more of an upward direction (and the greater the spin rate, the greater the lifting force), but when this backspin is mixed with sidespin, it creates the hacker's unintentional and dreaded slice or hook or the intentional fade or draw of a talented golfer.

Whereas dimples create lift and greater distance, a dimpleless ball generates the opposite: negative lift. A perfectly smooth golf ball spinning at the same number of revolutions per minute (rpm) that created upward deflection for the dimpled ball would experience deflection in the opposite direction. The principle of Dan and Fred's "Polara" ball, nicknamed the "Happy Non-Hooker," was really quite simple: Conventional dimples were retained on only about 50 percent of the surface and were concentrated in bands around the ball's equator (its seam), while around the pole regions were shallower dimples. In flight, the band of conventional dimples created the rough ball Magnus effect deflection—generating plenty of turbulence to create the lifting force to maximize carry. Meanwhile, at the poles, the smoother surface and shallower dimples created smooth ball Magnus effect deflection—deflection opposite the direction of the ball's spinning. Their design was a direct result of the paradoxical workings of the Magnus effect: the ability to simultaneously create a negative lift (deflection) to keep the ball moving in a straight line (horizontal stability) and to generate positive lift (deflection) from backspin to allow the ball to carry.

Fred and Dan were elated with their design. Its distance was equal to or better than that of other balls on the market, but the accuracy was far superior. The titans of golf ball manufacturing were caught by surprise by the Polara concept. Always obsessed with distance, they were asleep at the switch. They never considered the possibility that a better dimple design could not only improve distance, but also actually keep the ball on the fairway more often. (One prototype design was even able to mitigate a slice so effectively that the ball would overcorrect and veer left of center instead of slicing right.)

Before Fred and Dan introduced the Polara to the market, they sent manufacturing versions to the USGA for approval. At the time, the rule for a legal golf ball was simple and straightforward: "a ball must not weigh more than 1.62 ounces and not have a diameter less than 1.68 inches." The Polara golf ball met this standard, yet the USGA withheld approval. Fearing such improvements would "reduce the skill required to play golf and threaten the integrity of the game," an amendment against a nonuniform distribution of dimples was soon proposed. However, there was a major problem with this amendment: It seemed that almost all balls on the market had a nonuniform distribution of dimples. If they were going to institute this standard, only a couple of balls on the market would be legal.

The USGA eventually devised a peculiar rule to disqualify the Polara: "a golf ball must be spherical in shape and be designed to have equal aerodynamic properties and equal moments of inertia about any axis through its center." The test was implemented by striking a ball at two extremes (the equator vertical and equator horizontal) with a club-holding driving machine. Since the Polara had an intended asymmetrical pattern of dimple shapes and sizes, it showed slight differences in launch angle and maximum height for the two orientations. Because almost all balls on the market were asymmetrical for various reasons, they too showed differences in performance, but the USGA claimed their differences were within this new standard. The test for the standard was set at a value that would just rule out the Polara, and it became, in essence, the de facto "too good" standard.

• Dimples Here, There, and Everywhere

WHEN the early twentieth century golf pioneers found out that a nicked and cut ball traveled farther, which eventually led to the development of a dimpled ball, the rest of the world didn't pay much attention. In fact, it took over 80 years for the aerodynamic wonders of dimples to be widely understood and appreciated, which is why only now in the last few years the world has seen a plethora of innovative applications based largely on the dimpled golf ball design.

Bobsled drivers now routinely tack a small round piece of sandpaper to the front of their helmets to create the turbulent airflow that will more effectively hug the contours of the riders' heads. In baseball, Jeffrey DiTullo, an MIT Professor of Aeronautics, received a 1994 patent for a dimpled bat that he claims hitter's can swing 3 to 5 percent faster—a sizable enough difference to add 10 to 30 feet to long drives. And swimmers now use full bodysuits corrugated like sharkskin to reduce hydrodynamic drag.

It is important to remember, however, that dimples and other rough surfaces don't work for everything. For cylindrical objects of considerable length, such as planes, trains, and automobiles, airflow as streamlined as possible offers superior aerodynamics.

Fred and Dan sued the USGA, and showing no confidence in the defense that the ball "threatened the integrity of the game," the USGA wisely reached an out-of-court settlement with the inventors. A few years later a major golf ball manufacturer infringed on the Polara patents and had to reach an out-of-court settlement too. Because the patents have now long expired, the innovation trend begun by Fred and Dan has found a home in every golfer's bag. Nevertheless, longtime golfers should ponder the impact of dimple innovations on their game over the past few decades. If you were a big sheer in the 1970s and aren't now, you may be fooling yourself by attributing your improvement solely to a better swing or new clubs. Your drives may still have a considerable amount of slicing sidespin, but due to helpful dimple designs, it's no longer as exaggerated.

BUYING BALLS

There are over 90 different golf balls on the market, and the brand introductions are still continuing. Titleist, the largest manufacturer, goes so far as to manufacture 12 different balls. It would be tedious and confusing to try to give you detailed ball-by-ball comparisons; instead, it's more useful to offer some general guidelines in trying to select a ball that matches the way you play. You should think of balls falling into one of three camps:

1. **Two-piece balls** designed for recreational players offer durability, maximize distance, and minimize spin. The spin-minimizing ball is particularly advantageous if you are trying to cut down on your slice. This ball has a highly compressive acrylate or resin core and durable Surlyn cover, which gives it the greatest carry. It is popular not only because of its distance, but also because its cover is nearly indestructible. However, since these balls do not generate as much spin, it's more difficult to control, particularly for long approach shots when trying to bring the ball to a quick stop.

2. **Three-piece performance balls** for skilled players. You are sacrificing a little distance off the tee for more control around the green. Though not as durable as two-piece balls, they are more responsive—offering superior feel and control for those with middle to short game shot-making skills. But the trade-off is the need for greater skill in generating and controlling long-sailing drives with lower-lofted drivers, in particular, making sure the ball does not leave the clubface with excessive spin (pros use 7.5-degree to 8.5-

degree drivers with these balls). Cores have been made of metal, wood, and liquid-filled materials, and the covers come in either balata or Surlyn. Surlyn covers have been gaining in popularity because of its extreme toughness and the ability of manufacturers to produce covers that better mimic the feel and control of balata.

3. **Two-piece performance balls** that offer great length off the tee and recently have been able to match the feel, control, and shot-making characteristics of the best three-piece balls. This is possible because by making the cover thinner and the core larger ball designers can be more creative in mixing core materials to match seemingly competing objectives.

Because of better materials and improved manufacturing methods, there are excellent golf balls to be found coming from all three camps. The problem most companies have is producing balls that are good without being too good; in other words, not exceeding the USGA's Overall Distance Standard. Established in 1975, the guideline stipulates that to qualify for the USGA's list of conforming balls, the new models are subjected to a test, in which a robot hits a driver at a clubhead speed of 109 miles per hour, adjusting the collision until achieving a launch angle of 10 degrees and a backspin rate of 42 revolutions per second. Under these parameters, a ball passes if it travels 296.8 yards or less (including both carry and roll) under dry, windless conditions, and fails if it exceeds that distance. Since the optimal launch conditions for each ball (the best possible spin rate and launch angle) to achieve maximum distance are different, 10 degrees and a backspin rate of 42 revolutions per second is best for some balls and not others. Therefore, the USGA is in the process of using computer analysis (aerodynamic testing and "distance potential" testing data) to determine the optimum launch conditions that will maximize distance for each ball. They feel that testing balls with their unique distance-maximizing optimal launch conditions, rather than applying the same launch conditions to every ball, is the only way to come up with an Overall Distance Standard that is fair.

Within the confines of the Overall Distance Standard, manufacturers must try to cater to particular needs. Whereas recreational players show a strong preference for a maximum distance ball, Tour pros want a performance ball that balances length and control. Beginners and intermediate-level players should stick to the low-spin two-piece balls, as they achieve better distance off the tee and less slice or hook. Recreational players opting for two- and three-piece performance balls, such as the Maxfli Tour Patriot and Titleist Professional favored by the touring pros, do their game a disservice. These balls are much more responsive to spin, and using them, a typical recreational player ends up hitting too much backspin off the tee, and if they generate more sidespin than backspin at the point of contact, the ball will go much farther awry than with a low-spin two-piece ball. Moreover, recreational players do not have the shot-making skills to take advantage of the spin-friendly ball.

Tour pros sacrifice an insignificant amount of distance because they hit lower-lofted drivers that minimize the spin of their more spin-friendly balls, which is just another example of the subtle way that pros control the invisible forces. They know how to minimize spin for their drives, but also take advantage of the better compression to maximize it for the medium and short games—most importantly, the bite to hit and hold greens.

If you're an average recreational golfer and are not playing a course with difficult greens, you should stay with a durable and more accurate two-piece ball. But if you are a somewhat accomplished player who is fairly accurate and shoots in the 90s, using different balls for different courses and weather conditions might make sense. A performance ball is well suited to courses with very wide fairways and difficult greens, but for courses that feature narrow fairways and large, easy to approach greens, the low-spin two-piece ball is probably a better choice.

At the advanced level, golfers are interested in the relative compression of a ball. Compression is an indirect indicator of feel and control, giving some idea how much the ball deforms and hugs the clubhead before leaving. A few manufacturers have adopted compression rating systems as a way to convey feel; the greater the compression, the greater the feel. Of course, the problem for buyers is that there is no comparative standard: All the manufacturers have developed their own compression rating systems, making it impossible to compare different models.

Titleist is in a very enviable position because they command the lion's share of the market of three-piece and two-piece responsive balls used on Tour, regularly boasting that their balls win over half the PGA Tour events. This is not likely to change much even though new competitors keep entering the fray. Touring pros do not whimsically try new balls; after developing a feel for a certain model, they are hesitant to switch. Fortunately for the other manufacturers, more money is to be made in golf balls among recreational players. And now that professional and recreational golfers alike prefer the cheaper-to-manufacture two-piece balls, there is more money to be made than ever before.

You should demo a number of balls and select one that best matches your game, the courses you play, and the type of weather and golfing conditions you expect to encounter. If you honestly evaluate your skill level, you can buy balls that best match your needs.

LIMITING DISTANCE

The USGA messed up. Once you let the genie out of the bottle, it's hard to get him back in again. Legal and ever-better drivers and balls are resulting in greater distance off the tee for all golfers, so any rule introduced to take away distance will not be warmly received by most of the golf community.

Between the option of devising rule changes to limit the performance of the ball or the club, curtailing the sort of balls that

• A Cold Weather Secret

ALL golfers know you can't hit the ball as far as normal on a very cold day, but few know if they should blame cold weather or cold equipment. Werner and Greig tested equipment in cool (50°F), moderate (70°F), and hot conditions (100°F) and found that the decline or boost in distance due to air temperature is relatively minor. Compared to a moderate day, a drive will travel 3 yards farther on a hot day and 2 yards shorter on a cool day. What matters most is how temperature affects your equipment. It basically is about a decline in elastic energy: more energy gets dissipated in the collision as the temperature drops. It is the same reason why a field goal kicker and punter cannot kick the football as far on a cold day in December as on a warm one in September.

The club-ball collision doesn't feel the same when the weather is chilly; it doesn't even sound the same. You still can swing the club as fast as ever, but less of the energy is transferred from club to ball. That's because during the collision more of what should be elastic energy ends up as vibrational energy since the spring energy in the club and ball material itself are not working as well. And if your contact is off-center, you are going to feel a stinging sensation as your hands and arms painfully absorb the vibrational waves. There's always vibrational energy on impact, but on a cold day there is more of it, and it travels more quickly to the hands and arms. Better grips, perimeter weighting, and other dampening technology innovations have decreased the amount of vibrational energy that reaches the hands. Major League Baseball hitters, who swing the same ash bats they did a century ago, are not so lucky. Hitting a 95 mph fastball off the end of the bat on a frosty April afternoon is sure to result in a stinging pain shooting up the arms.

Although the decline in elasticity is detrimental for distance, it's far more problematic for shot-making. Your 100-yard

9-iron shot to the pin on a 100-degree day might be 20 yards short of the green on a frigid one. That is why a golf nut who likes to brave frosty conditions should consider ways to keep his clubs and balls warm. And between the two, it's far more important to keep the much more elastic golf ball warm.

The easiest and most obvious step is to store your equipment in a heated house. Unfortunately the benefit from this is rather limited, as playability will decline quickly in the cold air. Worse yet, it is really difficult to make adjustments to account for reduced elasticity as the temperature of your equipment declines from room temperature throughout the day. Since it takes over three hours to play a round of golf, storing several balls in a thermos of hot water and pulling out a fresh one for every hole is a good secondary option, and if you want to do something with your clubs as well, you can encase your golf bag in a small very well insulated DrumQuilt™. Designed to protect 55-gallon drums of temperature-sensitive cargo during shipping, a DrumQuilt allows your clubs to retain their ambient air temperature for a full day of golf, especially if you throw a couple of small hot water bottles under the quilt. Neither of these steps will match the feel of playing on an 80-degree summer day, and they are illegal if you're playing to compute your handicap, but it certainly will make playing far more enjoyable.

can be used is the easier and more effective route. Although a rule to limit the thinness of the clubface would put a cap on the coefficient of restitution (CoR) improvements related to the club, a lot more can be done to decrease ball distance. Unfortunately, the past actions of the USGA have made it difficult to introduce rules cutting back on ball distance. In particular, the USGA should never have set off on the course of creating complex testing and ever-changing standards to approve or disapprove the use of golf balls. The current test, the Overall Distance Standard, is another test that tries to create a "performance box" and is flawed because it tries to combine the testing of a ball's elasticity and aerodynamic properties into a single test instead of devising one test for each.

Because the USGA chose this course, they have been and will continue to be sued. Testing balls in this manner makes it difficult for manufacturers to design better ones that will still achieve USGA approval. For instance, if a manufacturer comes up with a new dimple design that decreases drag by 10 percent, it is unlikely that they could put that cover on their best core and windings because the ball will sail too far; the superior dimple design would require that it be matched with a less elastic (and therefore inferior) core. Moreover, because all golf balls are butting up against the edge of the "too good" distance box, it's difficult for manufacturers to distinguish their product lines from competitors or to design different products for different types of players. The USGA claims that their actions reflect a desire to protect the integrity of the game from technology, but by testing in this manner, it is neither fair to golf ball manufacturers nor people who play.

A much better system, and one that would save the USGA millions in legal defense costs, is to forget about any effort to control dimple design—including the pattern, depth, and shape—and instead solely test the CoR to set limits on elasticity by launching golf balls against a solid piece of titanium. Because the ball can be launched without spin during this test, it would judge a ball's elasticity (and a little bit of the dimples' minimization of drag), but not the dimples' ability to change the trajectory of the ball.

The crux of the distance boom stems from the latitude that the enlarged cores of today's two-piece ball allows manufacturers, making possible greater innovation in material mixing (enabling them to devise balls that achieve seemingly conflicting objectives), which has resulted in the CoR difference for high- and low-impact collisions to narrow. The CoR still declines at greater impact velocities, even with today's balls, yet the drop-off is less pronounced. In other words, today's ball plays about the same for short irons, but delivers more distance for woods and long irons and is far more rewarding to high velocity swingers. The result has been score deflation on the PGA Tour, which has forced tournament officials to react by lengthening par-4 and par-5 holes to curtail low scoring.

Limiting the CoR is technically feasible by simply altering the material properties of the core to change the energy transfer dynamics so that the ball dampens more energy and elastically returns less. If a CoR standard were instituted, there could be one allowable result, which could be skewed to more severely penalize very high velocity swingers. If you picture the core as having 10 distinct layers, this could be accomplished quite easily: Instead of all the layers having identical elastic and dampening properties, only the innermost layers could be more dampening—and therefore less accommodating to energy transfer. There would be no decline in the distance achieved by most golfers, only curtailing distance for those that swing 115 mph plus, who rely on the compression of the ball's innermost layers to achieve their greater distance. Most players never achieve the ball compression necessary to benefit from the elastic properties of the innermost material layers.

The greatest benefit of using a CoR standard would be the elimination of any criticism about the arbitrariness of the testing. Furthermore, it would objectively distinguish between what is legal and what is not. This is not a revolutionary approach: Baseball, another sport where there is always a worry of balls being too "juiced," uses this method. For an official Major League Baseball, the required CoR is 0.546 ± 0.032. The allowed for 0.032 variation means that the distance two identically hit baseballs travel can vary by 15 feet (4.5 meters). Because modern golf ball manufacturing and testing methods can ensure better consistency (that is, a more uniform CoR), the variation the USGA decides upon can be even smaller.

A maximum CoR rule would prevent super-juiced balls from

reaching the market, but still allow manufacturers greater latitude to tinker with playability characteristics. The only real threat to the integrity of the game is super-elastic balls that would turn tough par-4s and par-5s into easy birdie holes. A maximum CoR rule could eliminate the fear of this ever happening.

Leaving ball manufacturers the latitude to experiment with the ball surface—the shape, pattern, and depth of dimples—would not adversely affect the game. A dimple design specifically to limit the effects of a problem shared by many amateurs will always come with some trade-off that is unacceptable for the professional. For a severe slicer, a self-correcting ball, certainly keeps the ball nearer to the fairway, but renders it impossible to hit fades and draws. With this kind of trade-off, good players would never opt for a self-correcting ball, wanting one that generates sufficient backspin and lift to hold greens. When they hit a fade or a draw, they want the ball to do their bidding, not self-correct.

Allowing more possibilities for dimple design experimentation would increase the enjoyment of play for everyone. Beginning golfers using an error-minimizing ball do not have to spend as much time searching for errant balls because more would stay near the fairway, in turn speeding play, which benefits all golfers. Then, with improvement, beginners could move up to an intermediate-level ball that carries farther, holds greens better, and makes possible fades and draws.

The USGA should continue to allow manufacturers to develop different categories of balls to cater to the many different skill levels. In fact, they should even consider designating a still lower CoR ball so that short courses can play long in locations where the demand for golf is great and land is limited, and for courses at high altitudes and lower latitudes, where diminished gravitational acceleration adds 20 to 30 yards to a drive. For example, the USGA could impose a maximum CoR of 0.45 (with a 115 mph impact velocity) for these designated short courses. With a CoR of 0.45, 200-yard drives would be considered excellent. To address the concern of golfers who worry about losing their shot-making skills by alternating between short- and long-course balls, the ball can be designed so that it would retain the same 0.80 CoR (at 20 mph) for short irons as the long-course balls. In other words, the distance of a short-course ball would be curtailed off the tee and for long irons, but for short irons and around the green, the ball would play identically.

The ability of manufacturers to precisely mix materials makes the design of different CoR balls a practical option. Such a ball was once developed for the short courses found on the land-limited Cayman Islands but never caught on because they were lighter and badly affected by the wind. Wind problems would not be a worry for those developed today because reduced CoR balls can be manufactured at a weight comparable to standard balls.

• The Possibility of a New Game: Strategic Golf

BECAUSE the surface of the golf ball can be altered to achieve widely divergent flight patterns, it's possible to develop an intriguing alternative game much like disc (or Frisbee) golf, in which each player carries a bag of 10 to 15 discs, all with different aerodynamics. Depending on the circumstances, a different one can be used for every throw; the same can be done for golf. You can develop several different types of dimple designs for different circumstances such as one for exaggerated fades and hooks, a super-elastic one for driving, and a smooth ball for low trajectory "rollers" to stay under tree branches. There would be more strategy in this game as well because for every shot, you would be considering what type of ball to hit as well as what club to use and how to swing it. And short golf courses with an overabundance of trees, tight fairways, and very sharp doglegs in both directions could be designed specifically for this type of game.

The option of designing short courses that could be just as challenging as the best long ones would be warmly welcomed by golfers and course designers alike. The golf craze sweeping the world is creating Saturday morning traffic jams as the competition for tee times intensifies. Despite hundreds of new golf courses being built each year, the supply is insufficient to keep up with demand. These demographics have the USGA worried because a shortage of courses challenges the main plank of the USGA charter: "to make golf accessible to the masses." Sanctioning a short-course ball is the best long-term solution for creating affordable and challenging courses on limited real estate.

The Effect of New Equipment on Scoring

Despite the millions being spent each year developing better equipment, new gear did not appreciably improve scores until very recently. From the 1950s through the early '90s, winning scores on the PGA Tour did not dramatically improve, and if you compare golf to sports like swimming and track, this history seems all the more surprising. Record times have fallen and continue to fall in every major Olympic event; in fact, the world's best women swimmers today are recording better times than the top men of the 1950s. From the combination of better equipment, better-trained athletes, and better-maintained courses (e.g., greens, on average, are much faster—and consequently much truer rolling—than in years past), you would expect the average scores of professional golfers to have dropped just as much, if not more, but this simply did not happen.

Although useful statistics to measure amateur golfers to those of yesteryear do not exist, it is likely that the scores of beginners and intermediate players dramatically improved over the same period since the majority of innovations made clubs and balls more forgiving of errors. Equipment innovations have made the bad golfer not so bad, or stated slightly differently, bad golfers were able to become mediocre ones more quickly. If you give this some thought, it makes sense: Even when a pro misses a shot, usually it's only by a fraction, while the rest of us err more regularly and miss by much more. We're the ones who lack the smooth fluid swings to consistently strike the ball on-center. We're the ones who need help controlling slices and hooks. So we're the ones who most benefit from larger-headed drivers with enlarged sweet spots and balls more forgiving of errors, leading to better scores.

These dynamics began to change in the early 1990s as equipment design and material selection began to get more sophisticated. Innovations began to be fine-tuned to help the professional and low-handicapper as much as the high-handicapper. And some innovations, such as thin-faced drivers, actually deliver a far greater benefit to professional golfers who swing at very high velocities. Nevertheless, high-handicap golfers still have benefited more from innovation. Compared to a popular 1980 vintage driver, Werner and Greig found the modern driver has increased distance off the tee around 12 percent for professionals (a 250-yard drive has improved to a 280-yard drive), but a whopping 30 percent (a 137-yard drive translates to a 180-yard drive) for the high-handicap golfer. Moreover, while the professional golfer has experienced a modest improvement in accuracy due to better equipment, the high-handicapper has gotten significantly better accuracy.

Putters have also improved high-handicappers' scores more than they have the more accomplished golfers' tallies. The expanded sweet spot and longer dwell time of newer putters make it much easier for a novice to quickly develop a feel for putting. On the other hand, the low-handicapper does not need the enlarged sweet spot and has already developed a touch for putting that is only marginally enhanced by a putter's materials and design.

Better feel also has been provided by the improved frictional properties of clubfaces, allowing the ball to spin up the surface of the clubface at a faster rate, creating greater backspin. Although the typical high-handicapper can expect to generate 20 percent of his backspin from the friction of contact—as opposed to 80 percent coming from the angle of loft—the new clubface materials can increase backspin from surface friction another 5 to 10 percent. The low-handicapper, on the other hand, has always had a longer dwell time, so switching to clubs with better clubface surface material may only increase his ability to generate backspin by another 1 or 2 percent.

Confidence is an important piece of this puzzle. Pros develop a comfort zone with their older equipment; therefore it's a big gamble to switch to the most modern equipment even when the new innovations make them superior. This is particularly true when the equipment has a different feel, as any gain from better equipment may be offset by a problem with the adjustment: a lack of confidence in the motor memory to use the new equipment. That is why you hear of professionals willingly changing to the latest and greatest driver, their "power equipment," but showing much more reluctance to change short irons and putters, their "precision equipment."

To upgrade or not is a calculated gamble fraught with risk, especially for the more seasoned professional. It could be an abrupt career-ender either way: If an established pro is unable to adjust to the new equipment, confidence may fade and perhaps never return; on the other hand, it can be just as big a gamble not to upgrade if the up-and-coming young competitors, who grew up and developed comfort and confidence with the new gear, are gaining a significant advantage. In many ways, it's a "damned if you do, damned if you don't" proposition.

6

Injuries and Aging: *The Physics and Physiology Behind the Decline in Our Play*

IF golf is a precision game that does not require power or aerobic endurance, it would seem only logical that the best golfers in the world should stay that way for 20 to 30 years, but few do. Usually injuries and aging are to blame. Injuries and aging affect our games in more ways than you think. Besides the obvious physical toll, there exists also a mental toll. How we cope with not playing or scoring as well as we know we could have, or as well as we have in the past, is just as important as the actual physical decline itself.

Winning and improving are the two things that matter most to an avid golfer, and both are closely linked to mental and physical health. Unfortunately, regardless of your skill level, an injury or an age-related performance drop-off is the leading threat to put the kibosh on your improvement. Injuries and aging are probably the leading reasons that golfers give up on the game; unable to continue to play at a high level, it's easy to justify quitting. Despite the camaraderie that comes from enjoying a day out with friends, scores in the 90s are too depressing to accept if you were accustomed to scoring in the 80s. This is largely because the draw and enjoyment of any sport is to learn and gradually develop great aptitude for the game. It's only human nature to lose the hunger to play once the improvement stops. Runners are the same way: If a marathoner turned in a 3:10 time for this year's New York City Marathon, the motivation for next year will be to break the three-hour barrier. The same goes for dieting: It's easier to motivate to lose weight than maintain it once the targeted weight has been reached.

Obviously everybody's skills eventually wane; it's a physiological certainty. But age and injury need not curtail your prowess as much as you think. There's plenty that can be done to avoid injury and delay or minimize the age-related decline as long as possible. Much depends on the golfer's mental outlook: Whereas a depressive state will make things worse, a proactive outlook and an understanding of the nature of injuries and aging can greatly diminish the impact. The following sections explain the scientific reasons for injury- and aging-related performance declines and what preventative steps can be taken to avoid or delay the erosion.

Healthy Joints: The Key to a Healthy Game

The key to a healthy game is having healthy joints. Almost no one retires from golf because of muscle injuries, and the low impact nature of the game does not pose a threat to the skeletal system. If your physical skills decline, it's usually going to be because of joint problems.

Our joints give us mobility; it's useful to think of bones as levers and joints as the pivot points that control those levers. A lesser known, equally important, and subtler function of joints is that they provide stability. Effective execution of the golf swing depends upon the coordinated action of many joints, some primarily providing stability (such as the hip), others mobility (like the shoulder). Joint stability is primarily determined by ligaments, tendons, muscle tension, and the shape of the bone structure. Ligaments are the strong, predominately parallel-fibered bundles of collagen (connective tissue) that attach ends of bones together and control limb movement within a normal range, giving added stability to a joint; tendons, made of similar collagen fibers as ligaments, attach muscle to bone. Although tendons are stronger than ligaments, the tendons of muscles that overlap joints serve the same purpose as the ligaments: Any contractile force applied by the overlapping muscle generates a pulling force parallel to the long axis of the bone, thereby naturally increasing joint stability.

A narrow space called the joint cavity permits easy movement so that the muscles can power the joint, while still maintaining stability. The differences in shape and size of these cavities, as well as the joints themselves, have a bearing on stability too. For example, the deeper socket hip joint gives it greater inherent stability than the shallower socket at the shoulder joint.

Ligaments serve important flexibility and stability functions, but the muscles surrounding the joint are far more important. In essence, the muscles relieve much of the stress from the joints, and less muscle mass means that there is more stress left for the tendons and ligaments to absorb. Consider the elbow: Because the tendons are relatively small (compared to the muscles powering and stabilizing the joint) and tightly packed around the elbow joint, they must dampen many more times the vibrational energy than the more massive surrounding muscles. Logically, then, golfers with larger and stronger forearm muscles are better able to distribute the shearing force (from one segment sliding over another) originating above the elbow away from the tendons.

Rotational forces cause the most troubling repetitive stress for the joints. For example, at ball contact, a golfer's muscles are moving the club at the same time the ligaments and tendons at the elbow react to the shearing forces created at impact. These shearing forces are what result in tendonitis at the lead left elbow (for a right-handed golfer). It's considered a repetitive stress injury, the result of either many rounds of golf or several jarring vibrating wave jolts to the elbow from missing the sweet spot of the club.

Besides a single traumatic tear, ligaments and tendons can get permanently strained from repetitive stress that results in microtears, which is followed by inflammation. Classic examples of this syndrome are found among Major League Baseball pitchers: Few pitchers can throw as hard at 35 as they did a decade earlier. Because of strength training, almost all Major League pitchers these days test higher for strength at 35 than they did at 25, but usually their pitch velocity declines as the ligaments and tendons at the shoulder and elbow lose elasticity. (Every athlete is different, so besides a loss of elasticity, an overall decline in neuromuscular function, such as the muscle-nerve response time, can be a culprit as well). The same is true in football, as quarterbacks cannot throw as hard and far after many years under center, and a knee ligament injury for an NFL running back usually will take away some of his speed. When a golfer's tendons lose elasticity, they cannot as effectively store elastic energy during the backswing nor return it during the downswing. The result is a decline in the ability to accelerate the club as dramatically as was once possible.

Heredity plays a major role too: Genetics is not only a major factor in determining susceptibility to injury or an early decline due to aging, but also affects the way your body responds to training, practice, and play (the load and repetitiveness of the strain on joints). For instance, if two golfers followed an identical physical fitness program, it would not be uncommon for one golfer to show a 20 percent greater gain in strength and flexibility than the other.

Fortunately, for those more susceptible to injury, the nature of the game is in your favor. Golf does not entail the continual repetitive stress that causes the permanent strain of the tendons (like that resulting in "dead" arms for Major League pitchers). The significant amount of time between each stroke, coupled with the much lower strain levels of the swing (the impact being better distributed throughout the body), make golf a sport far less likely to permanently damage the tendons and ligaments. The one exception might be obsessed golfers who like to play 36 holes regardless of how hot and humid the conditions. The fatigue of walking 36 holes over eight hours and swinging a golf club up to 200 times can greatly heighten the chances of a repetitive strain injury. Taking that many swings in one day also increases the chances of one unbalanced, injury-causing swing.

An unbalanced or nonfluid swing creates uneven stresses that aren't well distributed throughout the body which can lead to an injury. The major problem usually occurs after ball impact, in trying to decelerate during the follow-through. If you don't have a smooth swing, it's a definite sign that there's too much pushing or pulling going on somewhere, too many of your muscles are fighting each other, with some joints subjected to much more of the deceleration torques than they should be. Although poor mechanics are often to blame, sometimes poor course conditions that result in loss of footing are the culprit.

Impact Injuries

When a sport is labeled "low impact" that means the activity entails less trauma for the joints. High impact sports like basketball, volleyball, and running can quickly begin to lose their appeal to the over-30 crowd because of the great physical toll, while contact sports like football and hockey are worse. They usually are given up even earlier, as athletes cry "no mas" to the physical pounding taken from collisions. But the sport of golf is different, and it can be played throughout a lifetime because of its low impact nature. Gravity-related forces are not continually jolting the body, which makes golf accessible to the extremely young as well as the very old. Walking is the major impact, but because it takes place over impact-absorbing turf, the stress is almost negligible.

The impact injuries that golfers are concerned with are specific to the wrist, elbow, and shoulder and usually occur when hitting through heavy rough, catching too much turf, or from hitting a root or rock hidden just below the surface. These swing injuries are most common among golfers with less skill, strength, knowledge, and instincts (as they are more unsuspecting of hazards just out of sight). Less-skilled swingers are most likely to be off target and miss the fairway altogether, which sets up the most challenging lies in the first place, while professionals, on the other hand, keep the ball on the fairway, confront fewer difficult lies, play better within their abilities (attempting fewer high-risk shots), and execute far fewer strokes per round (lessening the opportunity for fatigue-related injuries). But perhaps the major reason professionals avoid impact injuries is the ability to better brace, or "prestress," the joint when turf resistance is expected. Muscles have a built-in function of bracing the joint for an upcoming impact by exerting compression stress (a contracting force) on the part of the joint that's about to receive tension (a distending force) so that the amount of stress that can be absorbed by the joint increases proportionally. The bracing action of muscle is most apparent in downhill skiing because of the severity of the impacts involved. The countering muscle action called on is so extreme that when professional skiers go airborne, the compressive stress generated by their muscles sometimes continues and has been known to tear the anterior cruciate ligament (ACL). During flight, the countering gravitational forces no longer exist, yet the muscles continue to create compression forces, and the ACL tear occurs because the autonomous, preemptive muscle force on the joint remained in effect. Though the impacts involved in golf are far less traumatic, professional golfers, from years of kinesthetic memory playing difficult lies, are equally skilled at prestressing the joint. It also helps if you have greater muscle mass and strength because it dramatically augments the ability to create a bracing compression stress to withstand and absorb tension force on the joint.

For professional golfers, the greatest probability for injury occurs when they do not suspect an underlying root or rock. They are less likely to be prepared to prestress the joints as the hidden obstruction abruptly puts a stop to the swing, and the clubhead velocity slows from over 100 mph to a sudden stop in a fraction of a second. With no gradual deceleration, the shoulder, elbow, and wrist joints receive a jolting torque. Usually the wrist is most prone to injury since the torque of impact is first felt there, and the muscles that operate and stabilize the wrist are smaller and have less strength than those surrounding the elbow or shoulder. The wrists will absorb most of the impact torque well before any force at all is felt at the elbow or shoulder.

Some of this prestessing action involves firming up the grip at impact, but for good golfers, there's much more involved. As explained in Chapter 2, you only want a grip as firm as it has to be, and professional golfers spend a great deal of time studying a lie to accurately gauge what type of resistance to expect and what physical adjustments to make to account for the anticipated resistance as the club powers its way toward the ball. They make alterations to improve execution, but there's a subconscious and automatic process going on as well to gauge the amount of prestressing for injury prevention.

The Bad Back: The Most Feared Injury in Golf

The back is the key joint involved in the golf swing. A super limber back serves the critical function of being the middle link that "whips up" the momentum developed in the lower body that transfers to the upper body. When your back is not quite right, it's impossible to fluidly commute that force from the lower body to the upper body.

Unfortunately, back pain eventually afflicts almost everybody. Eight out of 10 Americans can expect to seek treatment for back pain at some time during their lifetime. Because golf is not a "back friendly" sport, golfers are even more likely than the rest of the population to sustain an injury. Back injuries account for up to 50 percent of all the injuries sustained by men who play golf. Worse yet, back injuries are usually debilitating ones, probably responsible for over 80 percent of setbacks that sideline golfers. You can continue to play with a little discomfort in your shoulder or wrist, but since the spinal twisting action of the golf swing puts tremendous stress on the back, the pain and awkwardness make it impossible to play.

Before considering the dynamics that cause back injuries, it is helpful to first look at the design of the spine, which acts much like a long series of short, connected rods—a chainlike mechanical structure—allowing the body a tremendous range of motion. Although no two vertebrae provide a large range of motion, collectively the 24 movable vertebrae permit an impressive range of motion.

One key element of a very flexible swing are the intervertebral disks. Located between vertebral bodies, they act as a ball bearing for loads requiring bending and twisting and as a shock absorber for compressive forces. Any stiffness in the intervertebral disks make a fluid and rhythmic swing impossible.

Golfers' back problems usually can be attributed to one or more of the following: excessive or uneven bending and twisting, "herky-jerky" mechanics, or muscular weakness. Of the three, muscular weakness is the most critical. Stronger muscles make it possible to do a far better job of stabilizing the spine: The stronger the muscles powering and supporting your spine, the more protected you are against the twisting or herky-jerky motions that cause injuries.

Few golfers realize that back-related injuries are more likely to be sustained during the follow-through after ball contact, which is the case because the clubhead is moving at 100 to 110 mph, and must come to a stop in about 0.20 seconds. Muscles to the left of the spine contribute significantly to swing velocity, but even before striking the ball, the muscles to the right of the spine are already prestressing the vertebrae to help absorb the decelerating forces necessary to bring the club to a stop during the follow-through. If the muscles are weak, they're not going to do a good job of contributing to the swing velocity or fluidly bringing the club to rest, which is especially true once fatigue has set in after playing three or four hours of golf. Thus, greater muscle strength and endurance augment the ability to absorb the decelerating forces of the swing.

Another aspect to be concerned about is relative muscular weakness. If golf is your only exercise, trunk muscles powering your swing will be much stronger than the supporting muscles decelerating the motion. Perhaps the only sport where strength asymmetry is worse is bowling, which is also the only sport that has a higher prevalence of back injuries than golf. Although unevenness in muscular strength is as bad or worse among the pros who do not engage in proper strength training, it's a much bigger problem for the rest of us because most pros' back and abdominal muscles are stronger than amateurs', so the strength asymmetry is not as disproportionate. That's why the personal trainers who work with professional golfers concentrate more on strengthening the underutilized support muscles than the primary powering muscles.

The peculiar twisting motion of the golf swing also contributes to back injuries, much in the same way that it does for bowling and snow shoveling. The popular belief is that snow shovelers hurt their backs from picking up too heavy a load of snow, but this is rarely the case, for even wet snow is not all that heavy. Instead it's primarily the combination of muscle atrophy from too sedentary a lifestyle and the twisting motion involved as the snow is lifted, swung around, and then thrown.

When the abdominal muscles are contracted, or tensed, they help unload and stabilize the spine from the twisting torque during the swing. Because the back muscles are forced into action with everyday activity while the abdominal muscles get virtually no workout at all if daily activity is limited to walking, standing, and sitting, they are going to be relatively weaker than the back muscles. The infamous and ubiquitous potbelly so many of us carry around does not help either, and definitely isn't a coveted physical trait for a golfer. An ideal physical profile for back injury prevention is the combination of stronger, tighter abdominal muscles, and more flexible back muscles. With better-conditioned abdominal muscles, the spine will be more effectively supported; with more flexible lower back muscles, the less likely a wrenching torque will cause a back injury.

If you already have back problems, there are two major biomechanical adjustments you can make to adapt your swing to account for an injury or restricted range of motion. First, narrow your stance to make it easier to execute a good shoulder and hip turn during the downswing and follow-through. Second, make sure that your hips turn freely throughout—closing them during the backswing and opening them during the downswing and follow-through. Starting with a more closed stance limits the amount of spinal twisting at the end of the backswing and start of the downswing, while finishing with more open hips curtails spinal twisting and helps better distribute the decelerating forces of the swing throughout the body, relieving much of the strain on the lower back. Aside from reducing the torque that the spine must absorb, these adjustments will make it easier to maintain good balance, rhythm, and timing.

Most professional golfers are acutely aware of the loss of back limberness as they age, but they have no idea why it happens. When asked about losing his competitiveness on the Senior PGA Tour, Lee Trevino replied "something happens to your back when you reach your mid-50s that no longer allows you to swing as smoothly as you once had." Age-related changes in the spine do indeed take place that make a smooth swing much more difficult. Intervertebral disks and the ligaments and tendons that support the spine degenerate with age, as does the bone structure. Moreover, the water content in the nucleus pulposus—the shock absorber between each vertebrae—decreases and progressively becomes more fibrous. In other words, the space between each intervertebral disk shrinks and starts to harden like concrete. This age-related tissue degeneration is the reason we start to lose some height as we age and feel a restriction in our range of motion and a decline in the fluidity of our twisting action. Trevino intuitively felt the kinks in his swing, a loss of the fluid rotation he for so long took for granted.

This begs the question: Are these age-related changes inevitable? The answer here is an equivocal one: Yes, eventually everyone experiences these effects of aging, but strength and flexibility training can limit or delay the decline in spine strength and flexibility as you age. Further, if you have already reached an advanced age and have never engaged in strength and flexibility training, studies have shown it's never too late to start. With the right strength and flexibility training, you can still dramatically improve your back muscle flexibility and abdominal muscle strength. Both will lower your risk of back injury and prolong your ability to maintain a high level of play well into your senior years.

- **Back Trouble**

Many of the greatest professionals have been sidelined by back problems. Jack Nicklaus, Lee Trevino, Fred Couples, Greg Norman, Tom Kite, Payne Stewart, Fuzzy Zoeller, Dan Pohl, Peter Jacobsen, and David Duval have all had their careers interrupted by back pain. Most responded to their injuries by consulting a physical therapist and embarking on a rigorous strength and flexibility program; more importantly, most wish they had dedicated more time to strength and flexibility training to prevent the injury from occurring in the first place.

Arthritis

Joints are surrounded by a capsule that protects them and holds the synovial fluid—a lubricant that allows the joints to move smoothly, similar to the way oil lubricates the moving parts of an engine. Synovial fluid lubricates and nourishes the areas of friction along the smooth-surfaced cartilage. Because adult cartilage lacks blood vessels and nerves, the back-and-forth flow of synovial fluid is vital to the survival of cartilage. Joints stiffen as the body loses some of its ability to lubricate them, and moreover, joint cartilage acts as the body's shock absorbers, so once it starts to wear down due to a lack of correct lubrication, we can experience painful bone-on-bone grinding, restricting our range of motion and increasing the likelihood of arthritic pain.

Besides a lubrication problem, stiff or painful joints can be the result of repetitive stress. Exercise that is extreme or creates an uneven stress can lead to the degenerative joint condition called osteoarthritis; the cartilage deteriorates or wears down, which seriously restricts movement at the joint. Osteoarthritis is also caused by diseases (diabetes), hormonal conditions (as in postmenopausal women), immune response, aging, or genetic predisposition. Around the age of 40 many people develop some twinges of osteoarthritis, and for an unfortunate few genetically predisposed to suffer from the degenerative joint disease, it flares up earlier and often with severely debilitating pain.

Since cartilage, unlike bone, is a tissue that cannot naturally repair itself, the remedies available to doctors have been few: They can either prescribe a steady diet of pain relievers or, in severe cases, resort to artificial joint replacement. However, there is tremendous hope that promising developmental drugs may eventually make it possible for everyone to enjoy pain-free golf well into their senior years.

The Mental Side of Injury

Many are familiar with the structural and physiological changes to the body that come from injuries and aging, but few people have the same knowledge of the psychological changes triggered by these biological effects that require adaptation. The physical recovery from an injury may take only a few weeks, but the mental readjustment may never happen.

Playing with stiffness, aches, and pain is a whole new experience, and a real threat to mess up your psyche. Golf is such a precision game that it only takes a little disturbance to throw off someone's game. We become less confident and focused because some of our focus shifts from the shot at hand to the pain—the sore back, shoulder, elbow, or wrist. Much of the time we end up altering the swing to take the stress off the injury. This alteration might be technically solid, but it also might mean that we have to redevelop kinesthetic memory for the touch involved in the newly adjusted swing.

An equally perplexing problem occurs once youVe recovered: You have to reestablish the confidence in your original mechanics. The brain needs to reacclimate itself to the preinjury swing but is likely to second-guess by considering the preinjury swing versus the experimental adjusted swing used while suffering the effects of the injury.

You have to feel good both physically and mentally to excel in golf. Thus, injury recovery involves not only the actual physical healing, but also renewing confidence in the body's ability to perform the instructions that the mind is communicating. It is extremely difficult to block out the kinesthetic memory of the altered swing, not to mention the fear of reinjury. Often a top professional will never again return to the top level even though he might be more physically fit than before the injury. Consider them casualties of confidence: They may be 100 percent physically sound again, but in their minds they think the injury is still holding them back; it's a phantom demon. The aches and pains are not there anymore, but the mind still thinks about the injury, which is a distraction certain to make execution tougher. The brain is still sending directions out to the peripheral nervous system, and the receptors in the joints and muscles are still receiving them, but memories of pain and stiffness make it hard to carry out the instructions and creates doubt about those commands.

Being Physically Fit to Hit

Back in the 1960s Gary Player began preaching about the virtues of physical fitness. At the time, his advice was largely ignored by fellow PGA professionals, and with good reason: Most of the successful players didn't sport great physiques. For those players who were not physically fit and who would rather not spend time on exercise, it was very easy and convenient to discount any claims of success attributed to good health. However, with the success of golfers like Tiger Woods and David Duval, who are dedicated to athletic training, popular opinion seems to be shifting. Nevertheless, there still remains doubters. Both Woods and Duval began playing golf as toddlers, so it's convenient and plausible to credit the greater plasticity of the very young nervous system to develop highly refined motor control with their success.

Even when great play is attributed to physical fitness, there is still a question of what aspects of it—diet, posture, stretching, weight lifting, plyometric training (muscle speed exercises), and aerobic exercise—are more important. All these elements of good health can help, but it's implausible to objectively rank each factor by relative importance. Thus, it's best not to gamble on one facet of athletic training while ignoring the others, and therefore every serious golfer should have a complete physical fitness plan that does not shortchange any of the various aspects.

Posture

Good posture is an underappreciated factor in preventing injury. Most of us probably are unaware of the fact that we are a few millimeters shorter at night before going to bed than in the morning. That's because after being subjected to the force of gravity over the course of a day, the spine compresses slightly. Likewise, astronauts who spend extended periods in a gravity-free environment usually grow 1 or 2 inches taller, with most of the lengthening of the spine coming from an expansion of the nucleus pulposus.

Poor posture exacerbates the effect of gravity's constant pull, so good posture is the only defense. A less than fully erect posture has the tendency to set off a cumulative decline, worsening spinal curvature over time. Further, poor posture can increase the amount of friction against nerves (as the intervertebral disks compress, "pinching" the nerve), resulting in a condition such as sciatica—a shooting pain that runs along the backs of the lower limbs.

Another reason to maintain good posture is to keep your game as uniform as possible. Just as we consistently go through the same visualization process and activate kinesthetic memory by using the same waggle, we want to keep our swing mechanics as constant as possible. Poor posture that leads to permanent pronounced curvature of the upper spine and more rounded shoulders, in effect, changes swing dynamics. The best way to ensure more consistent execution and a musculoskeletal system less prone to injury is to develop a greater awareness about maintaining a more erect posture with the shoulders back and the head up. It's all too common for those with poor posture to not even realize that they are not standing fully erect, which is why it is important that all golfers have their posture evaluated and continually work on improving it.

Recently yoga has been gaining in popularity among PGA Tour members as a means of improving posture as well as balance and flexibility. Yoga has also been cited as useful in helping to develop greater mind-body awareness, the same type of consciousness that best serves the golf swing. If the only benefit of integrating yoga into your training regime is a better awareness of your posture, it's still very worthwhile.

Strength and Flexibility Training

Stronger and more flexible muscles can improve your golf

game by augmenting your ability to generate increased clubhead velocity, to better brace against impact injuries, and to distribute the load to avoid repetitive-stress injuries.

For those who doubt the benefits of working as hard on your game off the course as on, picture Tiger Woods. He is one of the slighter-built men on the PGA Tour. Yet at the same time he is one of the bigger hitters, regularly driving the ball farther than 300 yards. Although a troubling paradox, it can be easily explained: Tiger credits much of his ability to whack the ball farther than almost everyone else to time spent in the Stanford University weight room working on his strength, quickness, and flexibility. As an 18-year-old freshman, he realized that he wasn't a very big guy, and wasn't going to be one. So if he was going to hit the ball like a bruiser, he better spend some time working out.

Of course there are quite a few people around the game of golf who do not believe in the virtues of off-the-course training. They rightly believe that rhythm and timing are far more important components of a sweet swing than strength; however, they wrongly believe that Tiger Woods's spectacular success—as well as the accomplishments of other true believers in training—is due solely to time spent on the practice tee. Unfortunately, many continue to perpetuate myths disparaging training.

Here are some of the more prevalent myths, followed by an explanation of why those beliefs are erroneous:

Myth #1: "Fm a strong guy and Fm very happy with my length off the tee, and because I practice and play a lot, strength training is a waste of my time."

To understand the benefits of strength training, you have to first understand how muscles work. Muscles act in pairs; they can contract but cannot elongate. In order to elongate, it takes a contraction from the muscle it's paired with. For example, a contraction of the triceps must follow a contraction of the biceps. If only our biceps could contract, we would be walking around with permanently bent elbows.

One of the best ways to improve performance is to repeatedly work the muscles used in your favorite sport or activity. Although playing a lot of golf ensures that the muscles powering your swing stay strong, those paired with them—the stabilizers of the joints—do not get enough of a workout from even a lot of golf. The lack of strength training partially explains the prevalence of shoulder injuries in golf as well as in baseball, tennis, and most other throwing and striking sports. That's because in any throwing or striking sport, the muscles used during deceleration—the "braking" muscles used during the follow-through of the swing—are relatively weaker. Water polo athletes are most susceptible to shoulder injuries, and with good reason: They throw the ball as fast as they can with their lower body submerged as the torso and arm move through the air. Because water is 70 times more dense than air, the upper body moves very freely through the air as the lower body is "held back" by the much denser water. Unable to move freely, the lower body cannot assist the upper body in decelerating the arm, which causes a tremendous destabilizing stress on the muscles that hold the shoulder joint in place. Much of the same dynamics occurs for golfers who swing awkwardly, slip a bit bringing the club around, or reach back for a little extra. The injury does not occur from generating swing velocity, but after contact when trying to decelerate the club from an awkward and unbalanced position. Since this can occasionally happen to even the best golfers, strengthening the stabilizing muscles to withstand greater stress by engaging in progressively more challenging exercises significantly lowers your injury risk.

Myth #2; "Strength training will ruin my game; a bodybuilder physique is too muscle-bound to swing smoothly"

Yes, it's true that strength training can make you muscle-bound. And yes, it's very unlikely that a professional bodybuilder could have a smooth golf swing—far too many muscles are impinging on each other as he brings his club around. However, the shared universe of bodybuilders and golfers is an extremely small one. Only a few bodybuilders golf; even fewer golfers body-build. In other words, your risk of becoming too muscle-bound to golf well is somewhere between slim and none.

The truth is that although Mr. Universe and Tiger Woods both lift weights, their goals—and thus their results—are completely different. To better understand this, it's useful to break down resistance exercises into their three major facets: the motion or the isolated muscle to be worked, the amount of weight to be used, and the number of repetitions. The diversity of results that can come about because of the right mix of these three variables is what makes strength training so beneficial for all athletes. A good weightlifting program uses a variety of resistance exercises to challenge the muscles to make them stronger; besides free weights and resistance machines, beneficial exercises can involve hoisting other types of heavy objects, lifting your own body weight, pulling against elastic bands, and pushing or pulling against immovable objects.

Bodybuilders want to add bulk all around, so they spend long hours doing numerous repetitions of exercises, working with weights equal roughly to 50 percent of the maximum amount they can lift once, which isn't what golfers should do. A golfer's strength training objective is to add as much strength as possible with little appreciable increase in bulkiness; in other words, the goal is to develop longer and stronger muscles. A golfer does more specific exercises (isolating specific muscle groups), fewer repetitions, and should lift about 75 percent of his maximum. And unlike a bodybuilder, who is only concerned with how his muscles look and not how they work, golfers need to combine flexibility exercises with strength training. If you regularly stretch, you will maintain or increase your range of motion as you gain strength, and stretching before physical activity helps to loosen up the muscles and increases range of motion. Both, in turn, decrease the risk of injury. However, it's important to remember that improper training also can overstretch joints. Like an overextended rubber band, joints subjected to constant loads may lose stability.

Although the way a joint moves is primarily determined by its structure, its range of motion is greatly influenced by several other factors including use, disease, injury, and the elasticity of muscles tendons, and ligaments. Stretching exercises lengthen muscles and help to maintain the integrity of the tendon and ligament tissues, thereby increasing range of motion in the direction of the lengthening. If a joint does not experience a full complement of movements, imbalances in range of motion occur. For instance, exercise that involves flexing, but never extension, will result in an increase in flexing, but a decrease in extension. That's why personal trainers emphasize increasing range of motion in the neglected direction to develop flexibility balance.

Regarding power, poor flexibility limits the velocity that the clubhead can attain. Flexibility exercises are designed to increase the range of motion of the muscles, allowing for a greater distance over which they can generate force, and consequently transfer greater momentum to the ball. Precision is affected as well because although we may be able to bring the club back the same distance before stretching as after, we cannot execute the backswing and start the downswing with the same fluidity. Prior to stretching we experience considerable tension at the 10 o'clock position, short of the angle constituting full range of motion; after stretching, resistance is more likely to occur near where we want maximum tension, near the 11 or 12 o'clock position. Lack of flexibility both unnaturally shortens the swing and makes it harder to control its path because the muscles are less capable of doing what the brain tells them to do. The right mix of stretching exercise ensures that you will not develop any kinks that will hinder your swing, adversely affecting your power and precision.

Improving your flexibility has the potential to do far more for your distance off the tee and your precision game than any regimen of strength training; you're not going to be fluid unless you're flexible. With almost every muscle in your body in action, either contracting or relaxing, you need the relaxing muscles to stretch out to their greatest length possible so that not only will the contraction of its paired muscles occur over the greatest length possible, but there won't be any kinks in the swing.

We can't all have a fluid and rhythmic swing because flexibility is, in large part, genetic, but there is plenty we can do to improve flexibility. All we have to do is work at it.

• The Secret of the Distance King

Golf's longest hitter is not a professional golfer, but a pharmacist. Jason Zuback, the winner of the RE/MAX North American Long Drive Contest in 1996, 1997, and the renamed RE/MAX World Long Drive Contest in 1998, regularly drives the ball over 350 yards and won the 1997 Championship by driving the ball 412 yards (the longest drive in championship history by 50 yards). Standing at 5-foot-9 and weighing 210 pounds, he attributes much of his superhuman driving ability to the strength training and stretching he does every day. Zuback, a former powerlifter, claims "overall strength is important, as long as you maintain flexibility." Yes, you need to improve both strength and flexibility. This doesn't mean that you too can match Jason's prowess—there is an undeniable genetic component to his driving talent—yet you can expect to notice significant improvements (Source: Golf Digest, *January 1999, p. 26).*

Myth #3: "Strength training is effective onlyfor young, athletic males who play sports like football, baseball, and basketball."

Strength training is good for anybody, at any age, and for any sport because exercise programs can be tailored for what you want to do. Not only can a strength training regimen be tailored to your age, but also can be designed to take into account how much golf you play. If you play every day, the emphasis should be on strengthening your underused muscles, but if you're only an occasional player, the focus should be more comprehensive, concentrated equally on the muscle groups that power the swing as well as the ones that stabilize the joints.

Heredity is also a major factor in determining your likelihood of sustaining injury, and your ability to keep up a high level of play well into your senior years. Genetics sets a probable range for predisposition to injury, so strength and flexibility training give you the opportunity to fulfill your predetermined potential.

Myth #4: "Whatever I gain in strength, I'll lose in touch."

Poppycock. There is no correlation between greater strength and less touch. In fact, the opposite may be true: Your touch may improve because of strength training, especially when combined with flexibility exercises. That's because the many swings involved in a round of golf will eventually fatigue some of your muscles, which lowers confidence because the brain starts to question its kinesthetic memory recall, meaning that you recalibrate every shot for "weariness" to account for a perceived difference in muscle strength. Strength exercises that entail two or three sets of 10 to 15 reps (with a weight equal to roughly 75 percent of your maximum capacity) will forestall the onset of fatigue and give you a more consistent feel or touch. You will have more confidence that the way you are stroking the ball on the 18th hole is the same as when you felt really fresh on the 3rd hole.

Myth #5: "If I lift weights, my hones and muscles will be tired and overused and consequently less able to perform well now, and in the long run it will shorten my years of good play."

Just the opposite is true; strength and flexibility training will lengthen your good playing days. Typically, muscles and their tendons surround a joint and—along with ligaments and cartilage—make it stronger. One of the most crucial objectives of athletic training is to increase the strength of the muscles that pass over joints, which improves joint stability, and thereby reduces the risk of sprains and dislocations.

The stronger the muscles surrounding your joints, the more stable they'll be. And because joint injuries are the ailments most commonly treated by orthopedic surgeons, strength training is recommended to lower the risk of injury. Likewise, the more flexible your musculoskeletal system, the more extended the deceleration period in the follow-through, and the more other muscles can get involved in assisting in the deceleration.

Myth #6: "Weight lifting builds bigger, slower muscles and I want smaller, faster ones."

Although strength training exercises can cause hypertrophy (enlargement) of muscle cells, there is no evidence that the muscles become slower as they get larger. In fact, it's more likely that the opposite is true, because when muscle is called upon to generate greater force, it responds by making more contractile proteins (the proteins inside muscle cells responsible for the cells' capacity to generate force). So the greater your muscle mass, the greater your potential to generate force.

However, as odd as it may sound, larger muscles and greater strength is no guarantee that you can accelerate a golf club any faster. A lack of strength by itself is rarely the culprit that limits the ability to generate clubhead velocity. Athletic training improves the *work* you can do, which is different than *power*. Strength is the ability of muscles to exert maximal force at a specified or deter mined velocity; power (work done/time) is about generating the explosive force needed to move quickly. Improving power entails heightening your ability to accelerate more quickly and explosively. This can be enhanced by plyometric exercises, which try to train your muscle fibers to swing faster, increasing your clubhead velocity beyond what is now your maximum speed.

Although scientific understanding of plyometric exercises is still in its infancy, for golf the reasons to employ this sort of workout are threefold: (1) to maximize the amount of elastic energy your muscles store at the transition from the backswing to the downswing; (2) to increase the speed and distance over which your muscles stretch; (3) and to improve the speed your muscles shorten (contract). In all, you want to get more muscle fibers to fire on cue, and for those already firing, to improve their ability to act more explosively.

Myth #7: "I don't have the time to lift regularly, so the greater muscle mass I achieve from strength training will turn to greater fat mass."

Muscle cannot turn to fat, nor fat to muscle; they are two different types of body tissue. What can happen, however, is that one can be traded for the other. If you stop strength training and at the same time do not lower your caloric intake to compensate for decreased activity, the extra calories will end up being stored as body fat. At the same time, you can expect your muscles to start to atrophy because of disuse. The concurrent increase in body fat and the decrease in muscle mass give the false impression that your muscles are turning into fat.

Myth #8: "I don't want to waste my money on training."

A common refrain is "I can't afford to get in shape—health clubs and exercise equipment simply cost too much." This is a short-sighted outlook because, in reality, you can't afford not to get in shape. A traumatic back injury sustained while golfing can lead to weeks of disability, followed by chronic pain over many years. After a bad back injury, many golfers never pick up a club again for fear of sustaining another injury.

Americans squander tremendous sums of money each year on exercise equipment, bought with good intentions, but which quickly end up collecting dust in a closet, basement, or attic. But the truth is you don't need fancy exercise equipment. You don't have to join a health club. You don't have to devote a lot of time. And you don't need to spend a lot of money. For less than $50, the cost of a set of dumbbells and a few other devices, you can develop a fairly effective strength training program that requires less than two hours each week.

Working with dumbbells is considered a dynamic or free weight exercise, and experts widely agree that free weight exercises are better than variable resistance machines for golfers because they can better mimic the actual movements used in a golf swing.

Aerobic Training

Physical fitness should not stop with strength training; you need aerobic exercise too. The cardiovascular and overall health benefits of aerobic exercise are well known, but few golfers realize the motor control benefits. More researchers than ever are convinced that regular physical activity does indeed delay the age-related changes in the nervous system by providing the stimulation to overcome the type of sensory deprivation that comes with the aging process. Aerobic exercise slows the loss of myelin (the nerve coating important for conduction), delays the selective atrophy of the brain that can cause mental confusion, and curtails the alteration in muscle function.

It's not easy to credit aerobic exercise with prolonging the careers of athletes in nonaerobic sports. Except for reaction time, it's hard to link a drop-off in performance to a decline in nervous system functions; it easily could be some other physiological factor instead. There is also the matter of genetic variables: aerobic exercise helps maintain everyone's nervous system, but genetics guarantees a wide variability of responses. In other words, everyone responds to aerobic exercise in a positive way, but the benefits are more appreciable for some than for others. What is known is that a mix of genetic and environmental factors influences performance decline and that inactivity accelerates the effects of aging.

It's unfortunate that many in the golfing community have bought into the myth that if you forgo the golf cart and walk the course you are getting aerobic exercise. Dr. Bill Mallon, the author of *The Golf Doctor* and the "Ask the Doctor" column in *Golf Digest,* has great disdain for carts since it "gives the golfer little aerobic or exercise benefit," which incorrectly implies that if you walk the course, there is an aerobic exercise benefit. This is nothing more than wishful thinking. You can burn plenty of calories and might improve leg muscle tone, but you are not getting aerobic exercise. Effective aerobic exercise requires an elevated heart rate (and greater oxygen uptake) for a minimum of thirty continuous minutes. The problem with golf as an aerobic exercise is the stop-and-go nature of walking, and even though it is four hours' worth, it is still not effective aerobic exercise.

Naysayers are likely to claim that strenuous and repetitive high-impact aerobic exercise wears out the joints, but this has proven to be false. From a lubrication perspective, aerobic exercise certainly benefits joints by elevating the pulse rate, which in turn increases the blood flow necessary for the distribution of nutrients throughout the body. A variety of all-around exercises is vital to ensure that our muscle and cartilage cells surrounding the joints remain well nourished; unlike bone, cartilage is a tissue of low metabolism, with limited regenerative capability. Suitable exercise thickens cartilage first by increasing its metabolism (and nutrition flow), and then by increasing collagen (flexibility cells) production as our joints attempt to adapt to increased stress. Surrounding ligaments and tendons work in conjunction with cartilage, so they also require a certain amount of stress and strain to maintain and increase strength.

It is believed that most joint injuries in noncontact sports result from a poor alignment between the joint surfaces, which is often hereditary, or are caused by a problem resulting from an earlier injury. As long as workouts are moderate and rest takes place when joints flare up with pain, we need not worry about a decline in the quality of joint tissue from aerobic exercise or from playing golf. In fact, if you incorporate an aerobic exercise like running into your training and you practice and play regularly, there is actually a little-known benefit to your skeletal system: As long as you slowly build up your practice schedule and have a well-balanced smooth swing, running and playing golf can improve bone density. The vibrational energy associated with clubhead-ball impact, or the impact from running, helps increase, respectively, your arms' and legs' ability to absorb calcium. Scientists believe that impact generates an electric potential in the bone that triggers a cellular response that increases bone remodeling. Your skeletal system adapts to these gradually increasing loads or repetitive stresses without damage because bone remodels itself in response to the manner and magnitude of the loads imposed on it. It's similar to the way skin tans when exposed to small doses of sunlight, but burns when exposed to strong sunlight for extended periods. Studies of athletes have confirmed adaptation and increases in bone density. One study found that the mineral content in the swinging arm of tennis players whose careers spanned several decades was, on average, 13 percent greater than the nonswinging arm.

By the same token, if you are sedentary all winter long and start golfing two rounds a day come April, you're asking for trouble. There's not enough time for the body to adapt to the repetitive loads. A golfer who hits 100 practice balls a day is far less likely to sustain a repetitive stress injury when practice goes up to 200 balls a day than for someone who starts hitting 200 balls a day after a six-month hiatus from any physical activity.

Aging and Inactivity: The Causality Question

The pace of aging varies with heredity, environment, and the level of physical activity. It's known that exercise, regardless of your age, can improve the replacement of worn-out cells, which is very beneficial to the bones, joints, and muscles. However, the nervous system is different. Some of the neural decline cannot be reversed since it involves the breakdown of cells that are not replaced. Greater activity cannot reverse this process.

The degree to which the biological processes of aging can be stopped or slowed down is unclear. We may never have a definitive answer about aging and inactivity. It is a chicken and egg question: Does aging lead to inactivity or does inactivity accelerate aging? There's also the problem of defining aging; we have a chronological age, a physical age, and a mental age. The former cannot be stopped, yet the latter two can be slowed by keeping the body and mind active. With a healthy lifestyle and training, it is possible for a 55-year-old professional golfer to show less physical and mental decline than a 40-year-old counterpart who did little to maintain overall physical fitness.

The face of competitive golf is changing, and undoubtedly we will see more young phenoms turning up on the Tour as more children are groomed for stardom and a more scientific approach is taken during their development. At the same time, we are going to see more competitive professionals over 40. The greater emphasis on a good diet, aerobic exercise, strength and flexibility training, and the wonders of modern medicine will make it possible for older players to compete far longer with their younger counterparts. In the not too distant future, it will cease to be a huge surprise when 45- to 50-year-old golfers more than occasionally win a Major. You need only look at some of the results of older athletes in other sports. Linford Christie was 32 years old when he won the 1992 Olympic gold medal in the 100-meter sprint and continued to win regularly until the age of 36; Merlene Ottey, a 41-year-old Jamaican sprinter, placed fourth in the 100-meter sprint at the 2000 Olympics; Michael Jordan was still the greatest basketball player in the world at the age of 35 (and returned to the NBA at 38); Nolan Ryan was one of the most dominant pitchers in baseball at the age of 45. All these athletes, aided by advances in sports medicine and the dedication to eat and train smart, remained among the best in the world in physically demanding sports.

Furthermore, several major championships have been won by golfers over 40: Jack Nicklaus won the Masters at 46, Julius Boros the PGA Championship at 48, and the U.S. Open has been captured by 45-year-old Hale Irwin and 43-year-old Raymond Floyd. But compared to the over-40 feats of athletes in other sports (Nolan Ryan in baseball, George Blanda in football, Gordy Howell in hockey, and Al Oeter in the discus), you would expect to see more 40+ year-old golf champions. Since golf is not a physically demanding sport—ifs primarily a game of motor control skill and not aerobic endurance and anaerobic power—the competitive age window should be much larger, especially when a Tour player is dedicated to physical fitness. All professional golfers have bought into the wisdom of maintaining a good diet and devoting ample time to strength and flexibility training, but very few engage in regular aerobic exercise, which is the last great frontier. To compete at a high level at more advanced ages, it's a necessity that the motor control functions stay sharp, and the benefits of aerobic exercise to the nervous system are too many to ignore.

When professional golfers start to concentrate on all aspects of physical fitness, it will undoubtedly prolong their careers. Older, more experienced players will always have greater wisdom and a better mental approach to the game, so if they put in the time to delay their physical decline and motivate themselves to practice as hard as ever and maintain the same burning desire to succeed they had in their youth, there is no reason that they cannot remain highly successful in their 40s and 50s.

There is little doubt that Jack Nicklaus could have won more tournaments if he engaged in aerobic exercise and strength and flexibility training throughout his career. Because Jack can't go back in time and relive his professional career, we will never know what could have been. However, it will be interesting to watch what will be; in particular, how well the young professional golfers who are physical fitness advocates will perform in their 40s and 50s.

7
Probability and Statistics

STATISTICS are about history. To learn from history and avoid repeating the mistakes of yesteryear, societies amass vast amounts of statistics from fields such as agriculture, commerce, transportation, and energy. These records provide valuable information about the past and present so that individuals, businesses, and governments can make informed, and often vital, decisions about the future. Sports are just another large harbor for statistics; and golf, like all the major sports, is a major generator. The end pages of golf magazines are regularly packed with numbers in an assortment of statistical categories.

Statistics are numbers that mean something, providing a way of thinking and questioning, a system that allows for extracting and interpreting useful information from those numbers. Analysts can present interesting, accurate, and informative statistics to present compelling connections, yet also employ them erroneously to sensationalize, inflate, and oversimplify. Often statistics are used selectively to embellish or demonize the play of a professional by a commentator starting with a premise such as "he chokes under pressure," and who then searches for the figures to back up the hypothesis. To create a compelling story, great liberties are taken in drawing conclusions based on the selective use of statistics even when the numbers, and the connections drawn, probably mean little, or nothing at all.

This chapter attempts to objectively look at some of the more intriguing questions of probability and statistics raised by golfers and fans of the game. The main goal is neither to find definitive answers nor demonize anyone, but to present how a statistician might look at the question at hand.

Comparisons in sports seem unavoidable in a culture that prizes excellence above all else. Although golf is not as prolific as baseball for generating numbers for statistical categories, there are still several to follow: yards per drive, putts per green in regulation, birdies per round, and scoring average are a few of the many categories that are monitored. A golfer's success or failure is usually attributed to some combination of these statistics. At the end of the year, evaluations are made such as "if the big and powerful John Daly had a respectable middle of the pack ranking in putts per greens in regulation, his superior power numbers would have resulted in at least a couple of victories this year."

Is there one statistical category that can predict the best player? Probably not: no results in any one statistical category can be pointed to as reason for success. Lowest score average is usually identified as the category with the greatest promise, yet in most years it's an unlikely predictor of who won the most majors, the most tournament victories, or even the most top-10 finishes. Although Tiger Woods recorded the all-time lowest strokes per round average of 67.79 in 2000 on his way to winning three Majors, and Lee Trevino was awarded the honor (represented by the Vardon Trophy) five times during a career that included six Majors, Jack Nicklaus never won the Vardon Trophy despite winning 18 Majors, and two in one year five different times. It turns out that sometimes the lowest strokes-per-round average is a very good indicator of how well a golfer fared in any given year, other times it is not. There probably will never be an accurate indicator since there are so many intangible factors that go into any tournament effort. Further, there's any number of random events that cannot be accounted for by any statistical measurement.

Lowest scoring average and other statistical categories may give us interesting numbers to ponder, but far less to draw compelling conclusions from. Probably the Tour players themselves are the most appreciative that the USGA compiles all these statistics; they serve as a window into the games of all their competitors, as well as a blueprint for improvement—what parts of their game need the most work. Undoubtedly professional golfers really don't need to look at their statistics to figure out what is ailing their game. They know all too well their strong points and deficiencies, yet these statistical categories give them a chance to gauge relative strengths and weaknesses. They can then put their limited practice time to the best use, spending the most time on the skills that call for the most work.

Tiger Woods: Greatest of All Time?

For golfing fans, winning remains the most popular and memorable statistic. Unlike team sports, greatness in individual sports is almost entirely focused on winning. The majority of baseball fans know who hit the most home runs or had the highest batting average even when that player's team ended up in last place; in comparison, few golf fans remember who topped the rankings for scoring average. They are much more likely to pay attention to who had the most tournament victories or who won the Majors. There is also the aspect of how those Major tournaments were won. Here opinions divide; some measure greatness by the ability to win by the greatest margin, while others put more weight on clutch performances—comebacks, winning when close, and the resolve to win in pressure situations. Still others measure greatness by long-term consistency and longevity, competing at a high level for 20-plus years.

In terms of labels for winners, the hierarchy for champions goes from great, to superstar, to hall of famer, to best of all time. The great label sticks after winning a couple Majors, superstar after winning some Majors and several other big tournaments in a short time span, and the hall of fame label is earned from impressive play for a decade or more. While there is some bickering about the worthiness of these labels when assigned to particular golfers, debate is most passionate when you try to anoint one stellar golfer as the greatest of all time. It's controversial since you're selecting only one golfer above all the rest despite the fact that many of the contenders never competed against each other, and when some of the candidates did, one might have been in his prime while the other was in the twilight of his career. It will never be easy to compare players who played in different eras and had different career lengths.

Nostalgic old-timers take refuge in the myths glorifying their ancient heroes, the likes of Bobby Jones, Sam Snead, Ben Hogan, Arnold Palmer, and Jack Nicklaus. They believe that the feats of these heroes of yesteryear far outshine those of their modern day counterparts. Jack Nicklaus was a legendary golfer in the 1960s, yet nostalgists point to the fact that he was still winning in the late 1980s, well beyond his prime, and are likely to explain away any statistics by attributing the superiority of modern day players to rule changes and improvements in technology. In particular, golf courses are better maintained and greens are cut shorter (being therefore faster and truer rolling), and lighter clubs and better golf balls make it possible to achieve greater distance and accuracy. Compared to 1980 vintage equipment, the typical 2000 model driver can be swung 3 to 5 mph faster, and the ball can travel 25 to 30 yards farther with considerably better accuracy. Better balls have added another 10 to 15 yards of distance, and also offer better precision.

Others argue nostalgists are looking at the world through rose-colored glasses. The old-timers never could have competed against the best golfers of today—the likes of Tiger Woods, Ernie Els, and David Duval—for several reasons: First, almost all professional golfers today began playing the game as toddlers and young children, when the nervous system has the greatest plasticity to adapt and "hard-wire" the intricate movements of the golf swing. Secondly, the golfers today are getting better and better instruction at earlier and earlier ages. Thirdly, the days of the out-of-shape professional golfer who always could be found at the 19th hole after finishing a round are over. Competition has resulted in better diets and a dedication to strength and flexibility training to improve results in the short-term and prolong good playing days for the long-term. And finally, the former champions' hagiographers fail to consider that athletes improve over time. Consider the world record times in all swimming and running events: In the early 1950s, people wondered if a runner would ever run a mile in less than four minutes. Fifty years later, not only has the four-minute barrier been broken, but the world record keeps falling, now approaching 3:45. Similarly, today's women swimmers regularly best most of the world records held by men in swimming in 1960. In all, better skill training, nutrition, conditioning, and opportunities give today's athletes an enormous advantage, ever improving performance and stretching human limits. Couple this with golfs growing worldwide popularity, the ever-expanding population pool, and the far more professional nature of the game (higher prize money, more tournaments, better practice facilities, and swing doctors) and it's indisputable that there are more good players today than ever before. The game is overall far more competitive today, yet it is far less clear whether the very best of today were better than the very best of yesteryear.

In reality, it's impossible to make a fair comparison between superstars in one era to those in another, though it never stops us from trying. You cannot even fairly compare the results of same-venue events because the courses evolve and conditions are never the same from year to year. Even a course like Augusta National, the home of the Masters, has undergone over a dozen major changes since the tournament began in 1934.

There is no argument that Tiger Woods, who won three Majors and nine tournaments in total, was the best golfer of 2000. But if he jumped on a time machine, could he have duplicated this feat in 1965 competing against the likes of Jack Nicklaus and Arnold Palmer? If Nicklaus broke into the PGA Tour today as a 25-year-old, would he win as many Majors? It's impossible to tell, and any conclusion can be easily discredited. It's easy to claim that the old-timers could not compete if they played today, yet to do so fails to take into account the intangible ability of all great champions, from any era, to rise up to the level of the competition. Moreover, unlike Woods whose childhood and adolescence were completely focused on golf, Nicklaus grew up as an athlete who also played high school football and baseball. This cross-training might have proven helpful to his eventual success in golf, yet it can also be argued that if he would have concentrated more of his time and energy to developing his golf game, there is no telling how many more Major tournaments he might have won.

An interesting way to compare players across different eras is to compare exceptional performance in any given Major tournament to all other superior exhibitions in that Major. Specifically, if Tiger Woods is indeed destined for the label "best of all time," how does his 15-stroke victory at the 2000 U.S. Open and 12-stroke victory at the 1997 Masters stack up against the results of other great champions? How does his exceptional performances in Majors compare to the others in the preeminent tournaments?

To make this comparison, you need to measure the number of standard deviations his victories were above the mean tournament total and compare that to the results of other champions. A standard deviation is a statistical tool that assigns a numerical value to any random outcome, in this case, the tournament total as it relates to the average (mean). It's the probable deviation from the mean of a random event, a statistical measure of variance, which is also called the sum of squares or the mean square. The variance is a measure of how far the outcomes are spread out. To come up with the variance, you first must determine the deviation of the sets of values from the mean. If the mean is 280, and there were only four outcomes (278, 273, 282, and 287), the deviations would be —2, —7, 2, and 7. Next square these quantities, leaving you with 4, 49, 4, and 49, respectively. The variance is the mean of these 4 quantities, and the standard deviation is the square root of that mean.

Standard Deviation of Four Outcomes

OUTCOMES	278	273	282	287
Deviations	-2	-7	2	7
Square quantities	4	49	4	49

Mean = 280

Variance = (4 + 49 + 4 + 49)/4 = 106/4 = 26.5

Standard Deviation = ^26.5 = 5.15

In Figure 7.1, a standard deviation can be visualized as a certain distance along the baseline of the curve from the mean or middle of the baseline out to the left or right to the point where the curve rises. Standard deviations break down so that about 68 percent of tournament totals fall within one standard deviation of the mean, 96 percent fall within two standard deviations of the mean, and 99+ percent fall within three standard deviations of the mean. Using this method to compare scores from different eras is a way to factor in the differences in conditions from year to year, changes in course layout and pin placements, and the differences in course difficulty (for example, the U.S. Open is not played at the same course every year).

Regardless of the Major, if a golfer scored two standard devia-

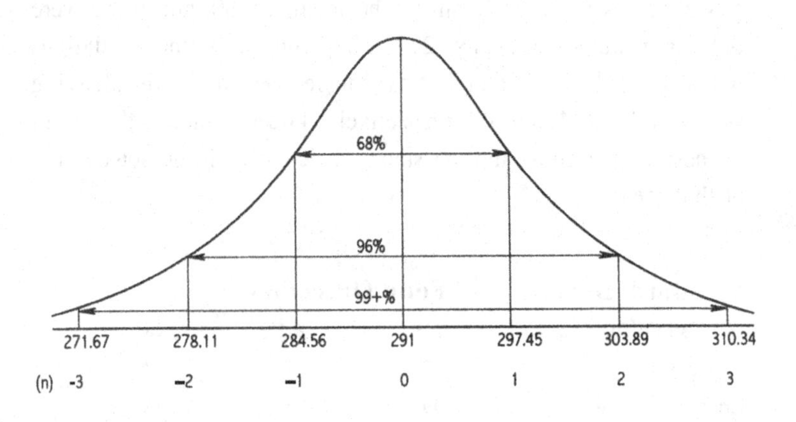

***Figure* 7. J:**
In the range of all scores recorded for a tournament, 68 percent fall within one standard deviation, 96 percent within two standard deviations, and 99+ percent within three. If the mean score for a tournament was 292, and the standard deviation was 7, then a score of 271 would be three standard deviations lower than the mean. For the 1997 Masters, Tiger Woods shot a 270, which was a remarkable three-plus standard deviations below the mean score of 291.

tions below the mean, he had a superb tournament. Scores in a Major that are three standard deviations below the mean are not only superb, but also so incredibly rare that it only happened three times in the 66 years that the Masters has been held. Table 1 shows the superexceptional talents who played phenomenally well at the Masters in its first 66 years. Ben Hogan's 274 in 1953 was the largest margin below the mean score (24 strokes), but the still large standard deviation of 9.481 resulted in his score being only 2.53 standard deviations below the mean. The 275 turned in by Fred Couples in 1992 was a score only nine strokes below the mean, but because of the very small standard deviation of 3.169, his number of standard deviations total was 2.84. Tiger's record 12-stroke margin and record low score of 270 was 3.26 standard deviations below the mean, yet both scores of 271 recorded in 1965 by Jack Nicklaus and 1976 by Ray Floyd resulted in higher standard deviations below the mean totals, 3.56 and 3.27, respectively.

It's interesting to note that Greg Norman, widely regarded as a tremendous talent, should have claimed a three-plus standard deviations tournament victory at the 1996 Masters. He was having an exceptional tournament (actually tying the course record with a first round 63) until his collapse in the final nine holes. If he just played well (exceptional wasn't required) over the back nine and finished with a 69 instead of a 78, his 272 would have been a fourth three-plus standard deviations performance (3.127). Had he continued playing exceptionally and matched Faldo's final round 67, his 270 total would have been 3.39 standard deviations above the mean—the second best of all time.

Table 1: The Masters

YEAR	WINNER	RUNNER-UP	MEAN*	STANDARD DEVIATION	NUMBER OF STANDARD DEVIATIONS BELOW THE MEAN
1935	282 G. Sarazen	282 C. Wood	297	6.607	2.57
1940	280 J. Demaret	284 L. Mangrum	296	5.496	2.91
1948	279 C. Harmon	284 C. Middlecoff	297	6.796	2.65
1949	282 S. Snead	285 J. Bulla, L. Mangrum	297	6.275	2.39
1951	280 B. Hogan	282 S. Riegel	297	6.823	2.49
1953	274 B. Hogan	279 E. Oliver	298	9.481	2.53
1955	279 C. Middlecoff	286 B. Hogan	300	7.535	2.55
1964	276 A. Palmer	282 J. Nicklaus, D. Marr	291	5.107	2.94
1965	271 J. Nicklaus	280 A. Palmer	292	5.895	3.56
1976	271 R. Floyd	279 B. Crenshaw	293	6.721	3.27
1980	275 S. Ballesteros	279 G. Gilbert, J. Newton	288	4.556	2.85
1992	275 F. Couples	277 R. Floyd	284	3.169	2.84
1995	274 B. Crenshaw	275 D. Love III	286	5.344	2.25
1996	276 N. Faldo	281 G. Norman	290	5.255	2.66
1997	270 T. Woods	282 T. Kite	291	6.445	3.26

(*includes finishes of 49th and better)

It can be justifiably argued that Jack Nicklaus in 1965 and Ray Floyd in 1976 turned in equally impressive performances as Woods's 1997 effort, but nothing compares to Tiger's 15-stroke victory at the hundredth U.S. Open at Pebble Beach, a four-round total of 272. Table 2 shows that Tiger's performance was an astonishing 4.75 standard deviations below the mean. Of all the five-stroke or better victories at the U.S. Open, the next closest performance was Ben Hogan's 1953 victory at the Oakmont Country Club (Oakmont, Pennsylvania)—probably the only other three-plus standard deviation U.S. Open performance.

Tiger Woods may indeed end up winning more Majors than Jack Nicklaus, but it isn't a sure thing. Yes, he is off to a quicker

Table 2: U.S. Open (five or more stroke victories)

YEAR	WINNER	RUNNER-UP	MEAN*	STANDARD DEVIATION	NUMBER OF STANDARD DEVIATIONS BELOW THE MEAN
1898	328 F. Herd	335 A. Smith	356	15.722	1.78
1899	315 W.Smith	326 G. Low, V. Fitzjohn, W. Way	331	9.563	2.51
1902	307 L. Auchterlonie	313 S.Gardner, W. Travis	334	11.616	2.32
1904	303 W. Anderson	308 G. Nicholls	329	12.385	2.10
1906	295 A. Smith	302 W. Smith	323	10.263	2.72
1953	283 B. Hogan	289 S. Snead	300	5.112	3.32
2000	272 T. Woods	287 M. Jimenez, E. Els	294	4.628	4.75

("Includes finishes of 50th and better)

start, having already won 6 Majors by the age of 25, yet whether he can sustain this type of performance over many years is questionable. Longevity is a major, albeit somewhat unfair, prerequisite for the title of Best of All Time. How do you compare the feats of a 10-year career to a 25-year career? Do 10 Majors in 10 years rank higher than 15 Majors in 25? Just as everyone has different God-given talents, unfortunate accidents (on and off the course), a career-ending injury, or a genetic predisposition to develop arthritis or age prematurely can obliterate a promising career. Take Seve Ballesteros: He was one of the best golfers of the 1980s, winning five Majors before the age of 30, but a debilitating back injury prematurely ended his career. Tiger Woods captured his first five Majors by the age of 25 and is more dedicated to maintaining and improving his physical fitness, yet a similar back injury can just as abruptly humble his aspirations.

A major advantage of most young successful golfers is a power game made possible by a super-limber back. However, as some of the flexibility declines with age, advances in the precision game have to offset some of the drop-off. While some power golfers like Nicklaus remained among the top ranks for years because of an always-improving precision game, others like Arnold Palmer faded. For example, Palmer won all seven of his Majors in seven years, from 1958 through 1964. As his power game declined, he could not offset that loss by making improvements in his precision shot-making. Tiger already has a better precision game than Palmer ever had, and his shot-making capabilities continue to improve, yet should he prematurely lose some of his super-limberness, it's highly unlikely he can dominate tournaments as he did at the 1997 Masters and 2000 U.S. Open. He still may win more than anybody else, but the chances of him winning three Majors again in one year after he reaches his mid 30s is highly improbable. Nonetheless, whatever happens in the future, Tiger always can be remembered as one of the greatest golfers of all time for the most exceptional performance at a U.S. Open and one of the top three showings at the Masters.

Probability

Whereas statistics are concerned with making inferences from samples to populations, probability works in the opposite direction: The population is known, so predictions can be made about obtaining various samples from the population.

Probability is the ratio of the number of times an outcome occurs to the total number of trials. All outcome probabilities must lie between 0 and 1, and the probabilities of all potential outcomes must sum to 1. An event expected to occur every time a trial is repeated has a probability of 1, indicating certainty; an event expected never to occur has a probability of 0, indicating impossibility, and most phenomena lie somewhere in-between, possessing a degree of randomness. When scientists use the term *random,* they do not mean the common definition: haphazard, accidental, without aim or direction. Instead they mean the inability to predict the outcome of individual events. In other words, carried out under identical circumstances, a random event may yield different outcomes. The most ubiquitous examples used in textbooks are coin tosses and dice throws; the outcome of each successive throw is completely independent of the previous toss. For instance, if a die comes up six 20 times in a row, the probability of getting a six on the next throw stays the same, as always: 1 in 6.

Because in sports it is nearly impossible to ensure victory (or loss) in any one game, outcomes are considered random events. The results of almost all sporting events entail varying degrees of randomness, though it's by no means equal. A game becomes less random the more the outcome depends on the talents of the players; while it's more so when the final score hinges on a single lucky bounce or any other factor beyond an athlete's control.

Golf is neither the most random game nor the least. The unpredictable ways golf balls fly, bounce, and roll, which are all highly dependent on wind, weather, and course conditions, add a significant amount of randomness to the game, but a sport like baseball is far more random for a number of reasons. First, baseball has several "all-or-nothing" aspects. The Major League Baseball batting average is around .250 (the probability of getting a hit is 1 in 4), which means batters succeed only one out of four tries. Individual outcomes are even more random, and the probability varies widely from one at bat to another, since it depends on the batter's skill as well as that of the pitcher. Further, victory depends solely on runs and not hits; the Cubs can have 14 singles in a game and not produce any runs, while the Braves can win the game with one "lucky" hit that does. Victory can be even more random since the Cubs can hit 14 additional well-tagged line drives that come to nothing because they were hit directly at fielders. These could be 14 textbook examples of good hitting, but none improve the chances of winning. Unlike the all or nothing nature of baseball, there's a lot of partial credit involved in golf.

When a shot is muffed and the ball carries 30 yards off course, it's not taking a direct route to the hole, but is still moving closer to the target.

Secondly, baseball is more random because the number of runs is small. If a baseball game could produce 270 to 290 runs, like the four-day total from a golf tournament, luck or chance would be much less of a factor. When a golfer wins a tournament by one stroke, say, 276 to 275, it can be credibly claimed that the outcome was decided by one "lucky" stroke—one friendly bounce or one putt dropping in instead of lipping out. The probability of a one-stroke victory is high, but this is more a reflection of the tremendous pool of competitors. Although the number-one and number-two golfers in the world may compete in 20 of the same tournaments in a year, the chances of one stroke separating them in any one event are remote.

If golf tournaments were only one round long instead of four, winning would be more random. The high scores compiled in a round of golf decrease randomness, and the four rounds that make up a tournament limit randomness even more. Often a relative unknown will take the lead in the first round and fade from the leader board over the next three days: The probability of maintaining top scores declines for less skilled players as more golf is played, and greater ability begins to shine through the higher the score totals. It comes down to the Law of Large Numbers—the more strokes that are made, the actual outcome will come closer and closer to the expected outcome of the best player prevailing. In other words, if the best golfer in the game was matched up against three other top players, the larger the sample, the closer the outcome will come to the expected probability. For example, if tournaments encompassed 10 rounds instead of four, Tiger Woods might have won 15 tournaments in 2000 instead of nine. Likewise, if you're a bettor and possess superior skill than your opponent, your probability of winning a wager is much greater when the outcome is based on a round of golf rather than a single hole.

Although baseball is more random overall, there are a few aspects of golf that are more difficult to predict. The most obvious is the field of play: Major League Baseball is played in over 30 ballparks, and while the dimensions and playing characteristics vary a bit from park to park, those relatively slight differences do not compare to the dramatic variations among the hundreds of tournament-caliber golf courses. Secondly, ballparks are maintained to prevent random bad bounces, while golf courses are designed and maintained to create hazards to collect the bad bounces. Even when you keep the ball where it's supposed to go, there's still a lot of randomness to how and where the ball will bounce and roll. Thirdly, wind and wet weather make matters more unpredictable. You never know if the 20-mph headwind will hold or if it will shift around to a quartering direction and gust at 30 mph when your ball is in the air. And you never know how much friction will be reduced by rain or what will happen to the wet ball hit from the soaked rough as it leaves the damp clubface. Will it roll with the near 5,000 rpm expected in dry conditions or at less than 2,000 rpm? Finally, there are both man-made and natural irregularities on the course. Besides the expected inconsistencies that come from divots and trampled greens, there are "invisible" natural irregularities. Identical input does not ensure identical output; in other words, no two putts will follow the same course. Using a putting machine, physicists Frank Werner and Richard Greig determined that the standard deviation for distance and lateral error is 2 percent of the total length. For a 15-foot putt, the standard deviation would be 3.6 inches (2 percent of 180 inches) in both lateral error and distance. This means that the putting stroke that dropped the ball dead center could be just as likely to roll past the left or right lip of the hole by over an inch. With this type of randomness involved, it's very understandable to watch all the body English employed by professionals to coax a putt toward the hole.

Limit Randomness to Improve Your Chances

Whatever the stroke—drive, long iron, short iron, chip, or putt—individual outcomes depend on varying degrees of skill and luck. Golf indeed takes great ability, yet there are several elements of randomness to the game where luck comes into play.

Luck is the product of a series of random events. Considering that you will be taking a great number of strokes a round, there's an appreciable amount of opportunity for good and bad random events to occur. Lucky is the fortunate bounce; unlucky the errant roll. A bad carom off the fringe, landing in a divot, or hitting a sprinkler head are random bad luck; a fortunate tree branch carom, a good bounce off the green's fringe, a ball sitting up in heavy rough, or a putt that lips in are random good luck. And at the end of the day, to do better than the expected probability is considered lucky; to do worse, unlucky. So when two athletes are evenly matched in every aspect imaginable, the fact is the outcome always comes down to a matter of luck: The winner gets the most random good bounces or the fewest bad ones. When one player is clearly better than another, he will always win in the long run: Superior skill almost invariably trumps a long string of uncanny good luck.

It might sound peculiar, but professional golfers do not dwell on the hope of luck or pray for their fortunes to turn around. All pros do things to take as much of the randomness out of the game as possible. They do this for a very simple reason: Luck is the effect of circumstances beyond control, and no one becomes a consistently good golfer unless they limit the potential for bad things to happen. The chances of good luck are just as great as bad, and you're as likely to get a good lie as a difficult one, a good bounce as a bad one, and a helpful roll as a discouraging one. Therefore, you should never let a string of unfortunate random outcomes get you down. As long as you stick to it, it's highly probable that a long string of bad luck with putts all lipping out will eventually be followed by a long string of good luck when they'll all find the cup. The better prepared you are mentally for the random ups and downs of the game, the better off you will be in the long run.

Top players do not square up to the ball thinking, T m feeling lucky, so I'm going to go for it." Instead they continually make decisions based on probability. They are always computing the margin of error for any given club selection or shot-making strategy. They might decide it's worth going for it, but this decision is usually taken because they feel that there's a high probability of success. However, even the best professional shot-makers have the tendency to either overrate their abilities or to err toward optimism about the probable outcome. Oftentimes they needlessly take risks.

Most of what was explained in Chapter 4, "Getting the Ball from Here to There," involved ways to improve your probable outcome. Always give yourself a large margin for error, so if a chip is off-line, the result won't be as bad. A pro plays a 7-iron shot with the probable outcome in the back of his mind that the shot may be off-line by half a degree; a hacker, on the other hand, should approach the same shot leaving a margin of error of something like 2 degrees. You also have to start asking yourself some simple questions: If I try to cream my drive in the hope of gaining 20 additional yards, is it worth the greater risk of an errant shot? Do I need to bring my ball right up to the water hazard to leave a short 40-yard approach shot, or should I give myself a greater margin for error and leave a slightly longer 60-yard approach? Unless you're a highly skilled golfer, giving yourself an extra margin for error almost always results in a better outcome. A ball hit from the rough with a 3-iron is probably going to land 20 yards shorter and with far less backspin than a 5-iron shot hit from 20 yards farther from the green but in the fairway. The same holds for water hazards; by laying up, giving yourself a larger margin for error for your tee shot, you only sacrifice a little distance that can be easily made up in the slightly longer approach shot. The risk of going for it is seldom worth the reward.

Another way to take luck out of the game is to limit bouncing and rolling. There are no unexpected bad bounces to be had in the air. You see the pros do this all the time; their high arcing hyper-backspin approach shots hit the green and either bounce and roll a few inches or dig in and roll backward. The ball isn't a victim to the vagaries of a bad bounce. For a professional's 7-iron that travels 140 yards, less than one yard of the total is bouncing and rolling; for a slow swinger's 7-iron shot that travels 103 yards, over 7 yards will be bouncing and rolling.

Of course, some situations call for the opposite strategy. If the pin placement is such that any ball that rolls up to the green will drift toward the cup, it makes sense to roll the ball up to the green to take advantage of the funnel-like pin placement. Pros go for the section of the green that gives them the largest margin of error, but try to impart the combination of backspin and sidespin that will bring the ball closer to the cup.

Except for very short approach shots, seldom do professional golfers hit the ball so that there is no kind of spin. By bouncing or rolling instead of staying airborne, lower fliers heighten the chances of a bad bounce or roll. Skittering across the unpredictable fairway or fringes of the green increases the chances of a random bad bounce. Unless there's a very gusty wind, the longer airborne flight and shorter roll of a backspin shot make it far easier for results to match intentions.

Putting is the one area of the game where randomness cannot be limited to any great extent. As long as the ball is bouncing and rolling, there's going to be random good and bad bounces. Moreover, if you're playing in the late afternoon, after hundreds of golfers have trampled the green taking some of the "trueness" out of the roll, you can expect more unpredictability.

In order to limit anxiety and consistently record as low a score as possible, golfers of all skill levels have to take as much of the randomness out of the game as their abilities allow. By incorporating a risk aversion strategy to your game—leaving yourself larger margins for error—you will be pleasantly surprised at the result. No longer will you have to endlessly pray for a dramatically lucky bounce or roll.

The Motivation to Play

Why do most people give up on the game? They hit the "wall," the point where anything they try—new equipment, lessons, practice, or the latest teaching gimmick—doesn't seem to help bring down their scores. Whether their wall is par, 80, 90, or 100, they cannot seem to get past it.

Regardless of talent, the law of diminishing returns tells us that there comes a point where the ability to improve our scores becomes more limited and ultimately must taper off entirely. However, the newer you are to the game and the higher your scores, the faster your improvement. A golfer who has a handicap of 35 will much more easily get it down to 30 than a counterpart trying to drop his 10 handicap to 5. That's because when so many of your shots are bad, just by limiting their number and severity it's easy to quickly bring down your score.

Breaking 80

On an early March morning in 1997, a vacationing President Clinton slipped off a step and injured his knee. Obviously, he was disappointed, but the focus of his frustration was particularly curious. "I felt real confident that today was the day I would finally break 80," he said, referring to his upcoming golf date with Greg Norman and the heightened confidence he had from the fact that The Shark—a top pro—was to be his playing partner.

This curious conviction that he would break 80 is not unique to President Clinton. Most avid golfers, regardless of ability, usually pursue an elusive, intoxicating goal that drives them to forsake family and friends to head for the links again and again. For the gifted, it's the dream to break par; for the duffer, it's the desire to break 80; and for the hacker, it's the humble goal of 90 or 100. Probably as a reflection of our goal-oriented society, no matter how poor or polished our skills, we cling to the belief that someday we will put together that truly exceptional round and reach that magical number.

More often than not, the reason for such confidence is only loosely attributed to improvements that come from extra practice, an insightful coaching tip, or even something as nebulous as good biorhythms. The main reason most duffers keep thinking "today is the day" has more to do with the optimistic way their minds compute probability. We tend to remember the times we played exceptionally well, and after playing a course 20 to 30 times, we will have played each hole well at least once or twice. Although her mean score for a given course may be 95, her combined best score for each hole over the last several rounds might be well under par. Few golfers keep all their score cards and make such calculations, but most remember stellar efforts and conveniently forget nightmarish ones. The selective memory of President Clinton, as well as that of most duffers, makes it plausible to see how golfers envision a truly exceptional day on which they shoot under 80.

In considering past scores, an optimist concludes that if on her most outstanding day—the one constructed from best-hole scores—she scores 68, a goal of 80 seems well within reason. The logic, however, fails to allow for external factors. For even the best golfers, it's a near physical impossibility to have perfect control over 80 shots in a row no matter how confident they may be. The slightest of errors at impact—the clubhead off-target by only 1 degree—can significantly alter the flight of the ball, and even professional golfers, many of whom practice for hours each day, consider it lucky to hit the fairway 18 tee shots in a row. Those exceptional rounds also almost always involve a considerable amount of good fortune, plenty of friendly lies and lucky bounces along the way.

With this in mind, Ted Jorgensen, a physicist and avid golfer, looked at the scorecards of a dreamy-eyed companion to determine his probability of achieving his long sought after goal of 80. For one par-4 hole, his scores over the last 22 rounds were two birdies (3 strokes), three pars (4), 14 bogies (5), two double bogies (6), and one triple bogey (7). Using these numbers, Jorgensen calculated the probability of his friend shooting a par on this one hole as 3 in 22 (0.136, or a 13.6 percent probability). This must be considered a very rough estimate because the sample is too small to satisfy the Law of Large Numbers. Nevertheless, the sample is large enough to reasonably project the long-term chances of this player breaking 80.

Jorgensen went on to calculate the score probabilities for the remaining 17 holes. Because the variability and range of scores for the other holes were quite similar, the chances of his companion breaking 80 turned out to be very slim indeed: less than 3 in 100,000. He would need to play one game a week for 2,000 years before he would be certain to break 80 three times. He could, of course, score 80 at any time, but given his past scores, such an extreme variation in scores is unlikely (assuming, of course, that his skill level did not improve appreciably).

Jorgensen didn't mention whether he conveyed the depressing probability to his companion. Perhaps it was better left unsaid, as nothing can squash a dream faster than the disheartening calculations of probability.

The Hole-in-One

To generate more excitement for a charity event, sometimes a luxury car is offered to anyone sinking a hole-in-one at a particular par-3 hole. Golfers who probably never thought about the possibilities of holing an ace all of a sudden feel that it's doable with a $40,000 car on the line. Perhaps this change in perception comes about because of the stakes, an unexplainable mind change akin to the way that many believe their chances of winning the lottery magically improve the week of a huge jackpot. This belief defies logic, and most will admit to such, yet almost all will claim an overriding feeling of good fortune. Couple this "lucky day" feeling with a heightened level of concentration, and the result is hundreds of true believers stepping up to the tee and liking their chances.

Strictly looking at the tee shot at hand—trying to get a 1.68-inch diameter ball to fall into an 4.25-inch diameter hole—the task seems daunting. At a distance of 160 yards, your allowable angle of lateral error (pushing the ball left or right of the optimum angle and still holing the shot) can be measured in the one one-thousandth of a degree range. For perspective, the margin for lateral error for a professional basketball player shooting a 15-foot free throw, which he sinks around 75 percent of the time, is 1.5 degrees.

If pinpoint on-line accuracy isn't a tough enough problem, you must also correctly figure out the right trajectory and velocity. For any given trajectory, there is a range of ball velocities that will result in an ace, and vice versa. A trajectory that is off by one degree or 1 mile per hour too little or too many can be the difference between an ace or the ball rolling to a stop 15 feet from the hole. Even when you defy the odds and the ball assumes an ideal direction, trajectory, and velocity, there's still a tremendous incalculable randomness that takes place once the ball lands. The layout and condition of the green around the hole—which determine the probability of a fortunate random bounce—are crucial. If the ball hits a bump, typically caused by a previous visitor's shoe spike, an on-line ball will veer off course. But by the same token, if an off-line ball hits a bump, it may change direction and bounce in the cup.

You must also take into account the huge variables of pin position and wind conditions. While a difficult pin placement and gusty winds will make the task nearly impossible, favorable placement and still air dramatically improve your chances. Gravity also can be a help or hindrance; as any avid golfer knows, there are certain large areas of some greens that will allow gravity to "funnel" all balls to a very specific location. If the pin is placed in the advantageous funneling area, the odds of a hole-in-one improve exponentially. It was just such a favorable pin placement that is partially responsible for the amazing feat of four golfers acing the same hole on the same day during the 1989 U.S. Open held at Oak Hill Country Club. On June 16 Doug Weaver, Mark Wiebe, Jerry Pate, and Nick Price all aced the 167-yard sixth hole.

Then just how improbable is a hole-in-one? From tracking tournament play, the USGA puts the odds of a professional golfer sinking a hole-in-one in a single round at 3,708 to 1. At these odds, a professional golfer who plays one round every day can expect to sink only one tee shot every 10 years. Further, the PGA estimates the odds of four golfers acing the same hole on the same day ever happening again at 332,000 to 1, which is still probably better than the odds facing a typical hacker trying to win the luxury car. Nevertheless, as hopeless as these odds seem, if s still possible to take comfort in the fact that the chances of a hole-in-one are still 37 times better than winning the New York State Lottery (12.3 million to 1).

Afterword

THOUGH you may be frustrated with your game, don't become disheartened; it's only natural. Golf can be a viciously difficult sport, and every golfer from weekend duffers to touring pros has at some point been disheartened by their play and is certain to be again. The degree of that despair, and how long it lasts, however, all depends on how you decide to deal with it.

The dynamics involved in golf are indeed complex, which is why an understanding of the underlying science can serve you well. It won't guarantee answers, but it certainly can give you the wisdom to ask the right questions. With your expanded knowledge, you can put puzzling problems in the correct perspective so that you can find—with the help of either a club pro, observant friend, or even a video camera—the solutions that best work for you. By taking a scientific approach to the game, it's much easier to develop a better swing, devise sound shot-making decisions, select clubs, and make adjustments for injuries and aging, all of which will lower your scores. Perhaps more important, your deeper appreciation for the subtle and invisible aspects of the game will not only better your results but also enhance your enjoyment of the sport. Good luck!

Bibliography

CHAPTER 2

Bernstein, B., et al. (1958) "On the Dynamics of the Bull Whip." *Journal of Acoustic Society of America* 30: 1112-1116.

Brancazio, P. (1984) *Sports Science.* New York: Simon & Schuster.

Cochran, A. and J. Stobbs (1968) *Search for the Perfect Swing.* New York: Lippincott.

Cochran, A. (ed.) (1990) *Science and Golf.* New York: Chapman and Hall.

Hay, J. (1985) *Biomechanics of Sports Techniques.* Englewood Cliffs, NJ: Prentice-Hall.

Jorgensen, T. (1994) *The Physics of Golf.* New York: American Institute of Physics.

Nicklaus, J. (1974) *Golf My Way.* New York: Simon & Schuster.

Lichtenberg, D. and J. Wills. (1978) "Maximizing the Range of the Shot Put." *American Journal of Physics* 46: 546-49.

Snead, S. (1997) *The Game I Love: Wisdom, Insight, and Instruction from Golfs Greatest Player.* New York: Ballantine.

Walker, J. (1975) "Physics of Karate Strikes." *American Journal of Physics* 43: 845-849.

Werner, F. and R. Greig. (2000) *How Golf Clubs Work and How to Optimize Their Designs.* Jackson Hole, WY: Origin Inc.

Zumerchik, J. (ed.) (1997) *Encyclopedia of Sports Science.* New York: Macmillan Library Reference USA.

CHAPTER 3

Annett, M. (1985) *Left, Right, Hand and Brain: The Right Shift Theory.* London: Erlbaum.

Buonomano, D. and M. Merzenich. (1998) "Cortical Plasticity: From Synapses to Maps." *Annual Review ofNeuroscience* 21: 149-186.

Coren, S. (1993) *Left-handedness: Causes and Consequences.* New York: Random House.

Gregg, J. (1987) *Vision in Sports.* Boston: Butterworths.

Hauger, B. and A. Rosenberg. (1997) "Motor Control," in the *Encyclopedia of Sports Science,* John Zumerchik (ed.), 700-719. New York: Macmillan Library Reference USA.

Lucas, J. and H. Lorayne. (1975) *The Memory Book.* New York: Ballantine.

Saunders, V. (December 1982) "How Vision Distortion Could Be Ruining Your Golf Game." *Golf World:* 54-56.

Snead, S. (1997) *The Game I Love: Wisdom, Insight, and Instruction from Golfs Greatest Player.* New York: Ballantine.

Spitzer, H. Desimone, R. and J. Moran. (1988) "Increased Attention Enhances Both Behavioral and Neuronal Performance." *Science* 240: 338-340.

Vander, A., et al. (1994) *Human Physiology,* 6th ed. New York: McGraw-Hill.

Watts, R. and T. Bahill. (1990) *Keep Your Eye on the Ball: The Science and Folklore of Baseball.* New York: Freeman.

Zieman, B. et al. (1993) "Optometric Trends in Sports Vision: Knowledge, Utilization, and Practitioner Role Expansion Potential." *Journal of the American Optometric Association* 64: 490-501.

Zumerchik, J. (ed.) (1997) *Encyclopedia of Sports Science.* New York: Macmillan Library Reference USA.

CHAPTER 4

Brancazio, P. (1981) "The Physics of Basketball." *American Journal of Physics* 49:356-365.

Brancazio, P. (1984) *Sports Science.* New York: Simon & Schuster.

Briggs, L. (1959) "Effect of Spin and Speed on the Lateral Deflection (Curve) of a Baseball." *American Journal of Physics* 27: 589-596.

Cochran A., and J. Stobbs. (1968) *Search for the Perfect Swing.* New York: Lippincott.

Cochran, A. (ed.) (1990) *Science and Golf.* New York: Chapman and Hall.

Davies, J. (1949) "The Aerodynamics of Golf Balls." *Journal of Applied Physics* 20: 821-828.

Erlichson, H. (1983) "Maximum Projectile Range with Drag and Lift, with Particular Application to Golf." *American Journal of Physics* 51: 357-362.

Hopkins, D. (1997) "Bowling," in the *Encyclopedia of Sports Science,* John Zumerchik (ed.), 85-107. New York: Macmillan Library Reference USA.

McDonald, W. (1991) "The Physics of the Drive in Golf." *American Journal of Physics* 59: 213-218.

McFarland, E. (1997) "Field Athletics: Jumping," in the *Encyclopedia of Sports Science,* John Zumerchik (ed.), 178-200. New York: Macmillan Library Reference USA.

Tan, A. (1997) "Basketball," in the *Encyclopedia of Sports Science,* John Zumerchik (ed.), 62-84. New York: Macmillan Library Reference USA.

Werner, F. and R. Greig (2000) *How Golf Clubs Really Work and How to Optimize Their Design.* Jackson Hole, WY: Origin Inc.

Williams, D. (1959) "Drag Forces on a Golf Ball in Flight and Its Practical Significance." *Quarterly Journal of Mechanical Applications of Mathematics* XII 3: 387-393.

Zander, J. and A. Chou (February 1999) "Max Out Your Ball: Increasing Your Launch Angle and Decreasing Your Spin Rate Will Help You Hit Farther." *Golf Digest* 50: 76-80.

Zumerchik, J. (1997) "Volleyball," in the *Encyclopedia of Sports Science,* John Zumerchik (ed.), 534-548. New York: Macmillan Library Reference USA.

CHAPTER 5

Anderson, J.D., Jr. (2000) "Aerodynamics," in the *Macmillan Encyclopedia of Energy,* John Zumerchik (ed.), 7-14. New York: Macmillan Reference USA.

Anderson, J.D., Jr. (1997) A *History of Aerodynamics, and Its Impact on Flying Machines.* New York: Cambridge University Press.

Brody, H. (1987) *Tennis Science for Tennis Players.* Philadelphia, PA: University of Pennsylvania Press.

Cochran, A. (ed.) (1990) *Science and Golf* New York: Chapman and Hall.

Davies, J. (1949) "The Aerodynamics of Golf Balls." *Journal of Applied Physics* 20: 821-828.

Easterling, K. (1993) *Advanced Materials for Sports Equipment.* New York: Chapman and Hall.

Holmstrom, F.E. and D.A. Nepala. Patent No. 3,819,190: *Golf Ball,* issued June 1974.

Martin, J. (1968) *The Curious History of the Golf Ball* New York: Horizon.

Millne, R. and J. Davies (1992) "The Role of the Shaft in the Golf Swing." *Journal of Biomechanics* 129: 975-983.

Notis, M. and D. Thomas (1997) "Equipment Materials," in the *Encyclopedia of Sports Science,* John Zumerchik (ed.), 153-177. New York: Macmillan Library Reference USA.

Werner, F. and R. Greig (2000) *How Golf Clubs Really Work and How to Optimize Their Design.* Jackson Hole, WY: Origin Inc.

CHAPTER 6

Adams, M. (1981) "The Effect of Posture on the Strength of the Lumbar Spine." *Engineering in Medicine.* 10: 199-205.

Adrian, M. and J. Cooper (1989) *The Biomechanics of Human Movement.* Indianapolis, IN: Benchmark.

Boone, T. and M. Foley (1997) "Aging and Performance," in the *Encyclopedia of Sports Science,* John Zumerchik (ed.), 583-601. New York: Macmillan Library Reference USA.

Christiansen, J. and J. Grzybowski. (1993) *Biology of Aging.* St. Louis: Mosby.

Cochran, A. and M. Farrally (eds.) (1998) *Science and Golf III: Proceedings of the World Scientific Congress of Golf* Champaign, IL: Human Kinetics.

Curry, J. (1984) *The Mechanical Adaption of Bones.* Princeton, NJ: Princeton University Press.

Huddleston, A. et al. (1980) "Bone Mass in Lifetime Tennis Players." *Journal of the American Medical Association* 244: 1107-1109.

Lind, D. (1996) *The Physics of Skiing.* New York: American Institute of Physics.

Mallon, B. with L. Dennis (1996) *The Golf Doctor: How to Play a Healthier, Better Round of Golf* New York: Macmillan.

McArdle, W. et al. (1991) *Exercise Physiology: Energy, Nutrition, and Human Performance.* Philadelphia: Lea and Febiger.

Mikesky, A. (1997) "Strength Training," in the *Encyclopedia of Sports Science,* John Zumerchik (ed.), 471-487. New York: Macmillan Library Reference USA.

Mundt, D. et al. (1993) "An Epidemiological Study of Sports and Weight Lifting as Possible Risk Factors for Herniated and Cervical Discs." *American Journal of Sports Medicine* 21: 854-863.

Spirduso, W. (1995) *Physical Dimensions of Aging.* Champaign, IL: Human Kinetics.

Stones, M. and A. Kozma. (1985) *Aging and Human Performance.* New York: Wiley.

Voight, M. and P. Draovitch (1991) "Plyometrics," in M. Albert (ed.) *Eccentric Muscle Training in Sports and Orthopedics.* New York: Churchill Livingston.

Westcott, W. (1995) *Strength Fitness,* 4th edition. Dubuque, IA: Brown.

Zumerchik, J. (1997) "Skeletal System," in the *Encyclopedia of Sports Science,* John Zumerchik (ed.), 842-869. New York: Macmillan Library Reference USA.

CHAPTER 7

Armenti, A. (1992) *The Physics of Sports.* New York: American Institute of Physics.

Fischer, G. (1980) "Exercises in Probability and Statistics, or the Probability of Winning at Tennis." *American Journal of Physics* 48: 14-19.

Gould, S. (1986) "Entropic Homogeneity Isn't Why No One Hits .400 Anymore." *Discover* (August): 60-66.

Jaffe, A. (1987) *Misused Statistics: Straight Talk for Twisted Numbers.* New York: Dekker.

Jorgensen, T. (1994) *The Physics of Golf* New York: American Institute of Physics.

Tan, A. (1988-89) "Athletic Performance Trends in the Olympics." *Mathematical Spectrum* 21:78-84.

Tan, A. and J. Zumerchik (1997) "Statistics," in the *Encyclopedia of Sports Science,* John Zumerchik (ed.), 453-470. New York: Macmillan Library Reference USA.

Tan, A. and J. Zumerchik (1997) "Basketball," in the *Encyclopedia of Sports Science,* John Zumerchik (ed.), 62-84. New York: Macmillan Library Reference USA.

Watts, R.G. and A.T. Bahill (2000) *Keep Your Eye on the Ball: Curveballs, Knuckleballs and Fallacies of Baseball.* New York: W.H. Freeman Company.

Index

About the Author

JOHN ZUMERCHIK has had a rather varied career as an editor/writer. He was an editor for the *Encyclopedia of Sports Science,* a two-volume reference work covering the physiology and physics of sports as well as the physics of sports injuries. (Though listed as editor, he authored or rewrote over half the entries.) He also served the dual role of author/editor for the three-volume *Macmillan Encyclopedia of Energy*. He previously served as a senior editor of the American Institute of Physics, and as a physics, forestry, and geology editor for the College Division of McGraw-Hill. He lives in New York.